7th European Workshop on Structural Health Monitoring (EWSHM 2014)

AA005236

Conference Program and Abstract Book

Nantes, France
8 - 11 July 2014

ISBN: 978-1-5108-3203-9

Printed from e-media with permission by:

Curran Associates, Inc.
57 Morehouse Lane
Red Hook, NY 12571

Some format issues inherent in the e-media version may also appear in this print version.

Copyright© (2014) by INRIA Rennes – Bretagne Atlantique
All rights reserved.

Printed by Curran Associates, Inc. (2016)

For permission requests, please contact INRIA Rennes – Bretagne Atlantique
at the address below.

INRIA Rennes – Bretagne Atlantique
Domaine de Voluceau
Rocquencourt – BP 105
78153 Le Chesnay Cedex
France

Phone: +33 (0)1 39 63 55 11
Fax: +33 (0)1 39 63 53 30

www.inria.fr

Additional copies of this publication are available from:

Curran Associates, Inc.
57 Morehouse Lane
Red Hook, NY 12571 USA
Phone: 845-758-0400
Fax: 845-758-2633
Email: curran@proceedings.com
Web: www.proceedings.com

airbus.com
The A350 XWB.
Next generation
efficiency with
next level comfort.
• 25% lower operating costs than previous generation competitors
• All-new composite fuselage, highest operational reliability and outstanding aerodynamics
• Exceptional level of comfort with more personal space, wider seats and panoramic windows, inflight entertainment systems with full connectivity
Become part of the A350 XWB family, the most successful widebody aircraft of recent years.
© AIRBUS, 2014. All rights reserved. Airbus, its logo and the product names are registered trademarks.
AIRBUS

EWSHM 2014

July 8-11, 2014

Nantes, France

www.ewshm2014.com

IFSTTAR

informatics mathematics

UNIVERSITÉ DE NANTES

IRT
JULES
VERNE

A mutualized research centre for advanced manufacturing

Since 2012, IRT Jules Verne, a mutualized industrial research institute dedicated to advanced manufacturing, brings together manufacturers, training facilities, private and public applied research laboratories, and prototyping and industrial demonstration resources. With a co-built and shared strategy, it runs its own technological research activity organized in project mode and benefits of skills and equipments co-location at a single site in Nantes. Its core research involves developing key enabling technologies at the intermediate TRLs (4 to 6) in order to develop competences such as: design, simulation and production of large scale and complex structures; design and fabrication of multi-material assemblies (composite, metallic and hybrid structures); structure performance and process simulation; robotics and cobotics for industrial processes; recycling and waste upgrading of composites

One of the IRT's research project : Robofin about a new robotic solution to finish large size parts

32 industrials

- An R&T budget planned of 350 M€ by 2020
- 30 projects already launched for a total amount of 40 M€
- 6 technological platforms with skills co-location fitted with large scale R&T equipment

3 major axis

- Integrated product/process design
- Innovation processes
- Adaptive and intelligent manufacturing systems

www.irt-jules-verne.fr

Sébastien Leroy
Business Development Manager
sebastien.leroy@irt-jules-verne.fr
Tél. : + 33 (0)2 28 44 36 11

CORPORATE SPONSOR

AIRBUS

Sponsoring the gala dinner (Wednesday, July 9)

GOLD SPONSORS

IRT JULES VERNE

Structural Vibration Solutions A/S
SVS

Optilab

Optilab

Optilab, a US-based photonics technology company, designs, manufactures and markets innovative and cost-effective laser and photonic solutions to its customer's systems. The innovative products include FBG Sensor Interrogator, FBG Analyzer, and FBG's Qualification. The FBG sensor interrogators span three models. FSI-OM is a compact, rugged, dynamic interrogator module designed for demanding aerospace applications while FSI-OD is a weather resistant, water proof interrogator system designed for demanding outdoor applications. The indoor, single 1U applications are carried out by FSI-RM.

The recent expansion in research and development facilities marks a milestone for Optilab. Optilab owns and operates OEQuest.com, one of the largest web portals in the photonics industry.

SMARTEC

Smartec

Since 1996, SMARTEC is leader in the application of Fiber Optic Sensing for Structural Health Monitoring. Our own products and solutions are based on SOFO, Fiber Bragg Grating, Fabry-Perot, as well as Distributed Brillouin and Raman sensing technologies.

Besides its complete range of off-the-shelf fiber optic sensors, SMARTEC offers Engineering Services designed to support you in the design, implementation and management of simple to very complex monitoring systems.

SMARTEC is part of Nova Metrix, the world's leading provider of Geotechnical, Environmental, and Materials Testing products serving the Structural Health, Construction, Civil Engineering and Energy markets. Nova Metrix Brands also include Roctest, FISO, Telemac, Sensornet, Durham Geo Slope Indicator and Sherborne Sensors.

EWSHM 2014 IS SPONSORED BY

AIRBUS

IRT JULES VERNE

Structural Vibration Solutions A/S SVS

SMARTEC

Optilab

EWSHM 2014 IS SUPPORTED BY

Nantes Métropole COMMUNAUTÉ URBAINE

Région PAYS DE LA LOIRE

precend

MRGenCi

Evaluation et Contrôle Non Destructifs ECND PAYS DE LA LOIRE

EMC2

SAGE

HAL archives-ouvertes

NDT net

IFSTTAR

Inria informatics mathematics

UNIVERSITÉ DE NANTES

Contents

Conference Overview

After years of being an emerging topic, SHM has now matured into a solution for addressing the challenges of safe structures (aircrafts, civil engineering constructions...). SHM deals with both life-cycle monitoring and early detection and quantification of damage. As an illustration, since 2003, the European Community has spent more money on Civil Engineering maintenance than on the construction of new structures.

For most SHM applications, it appears that no single method or technology can be the one and only solution for any problem, while the solution generally resides in the relevant use of a large spectrum of techniques. This is reflected in the rich and inhomogeneous aspect of our community, composed of experts in many different fields. Our common challenge is to identify and investigate the convergence that paves the ground for Smart Structures. Moreover the resilient theoretical questions or development of technologies should be addressed to improve measurements, data processing, decision methods and aid tools. The 7th European Workshop on Structural Health Monitoring 2014 is devoted to address these challenges from material to structure, from the actual data and models to the forecasting of behavior.

The 7th European Workshop on Structural Health Monitoring 2014 organized in Nantes represents a new step in the SHM history whose success is confirmed each year. Organized since 2002 in Europe, alternately with IWSHM (held in Stanford University, California) every odd year, the success of EWSHM has significantly increased: number of attendees, company presence (through exhibition booths, visits...), multidisciplinary links with other communities (e.g. PHM) or application fields. The European Workshop on Structural Health Monitoring has been organized in Cachan/France (2002), in Munich/Germany (2004), in Granada/Spain (2006), in Krakow/Poland (2008), in Sorrento/Italy (2010) and in Dresden/Germany (2012).

The co-location of EWSHM and PHM (Prognostics and Health Management) conferences in Dresden in 2012 is renewed and extended in 2014. Participants of each community have free cross-access to both SHM and PHM parallel sessions; a common SHM/PHM keynote lecture session is organized as well as the "in action" plenary session is organized for and by PHM and SHM members.

IFSTTAR, Inria and Université de Nantes, organizers of EWSHM 2014, have paid special attention on opening the conference to other applications fields that present promising developments for SHM: Railways, Wind Energy, Nano technologies... The link with the economic and industrial world is also one of the organizer's objectives: a full exhibition area is devoted to companies linked with SHM, visits of Airbus as well as operational laboratories such as Jules Verne and IFSTTAR institutes are also proposed.

Symposium Chairs

Vincent Le Cam (IFSTTAR, France)

Vincent Le Cam is research-engineer at IFSTTAR (French Institute of Science and Technology for Transport Development and Networks). He graduated from Polytech Nantes engineering school in 1997 specialized in electronic. He has been working 4 years in Alcatel-Lucent company as engineer, team manager and consultant. In 2002, he graduated from ENTPE (French National Engineering School of Public Works). Now, at IFSTTAR, he is working as co-head of SII (Structures and Integrated Instrumentation) Laboratory (25 people). His interests are smart systems, sensors and wireless sensor networks in view of coupling structural modeling to smart devices. The topics of his 25 peer reviewed journal and conference papers refer to Non Destructive Techniques (NDT) and Structural Health Monitoring (SHM). He is currently member of the joint team between Inria and IFSTTAR dedicated to SHM and member of EWSHM International Scientific Committee.

Laurent Mevel (Inria, France)

Laurent Mevel is researcher at Inria (French National Institute for Research in Computer Science and Automatic Control). He graduated from University of Rennes in 1997, where he received a Ph.D. in Applied Mathematics. Now he is working on identification and detection in the context of modal analysis of structures under natural excitation. His interests include: structural health monitoring and detection for vibration mechanics, modeling and rejection of environmental effects, model validation and uncertainty estimations and model reduction of large systems. He was involved in 5 European projects. He directed 8 theses on these topics. He is the author or co-author of 34 journal papers and more than 130 international conference papers. He is currently team leader of I4S team at Inria, a joint team between Inria and IFSTTAR dedicated to SHM.

Franck Schoefs (Université de Nantes, France)

Franck Schoefs is full time Professor at Université de Nantes, GeM (Institute for Research in Civil and Mechanical Engineering, UMR CNRS 6183) / IUML (Sea and Litoral Research Institute, FR CNRS 3473). He got his PhD on Structural Reliability in 1996 and is the leader of the group « Structural Monitoring and Computation Group » (35 people) since 2010. He took the head of "Marine systems, structures and geomaterials" at IUML in 2012. He is director of the Master of Science in Civil Engineering at Université de Nantes (120 students). He was involved in 4 European projects and is Scientific Officer of 20 contracts in 15 years and co-supervisor or director of 12 theses. He is director and member of the Scientific Council of Scientific Interest Group "Risk Management and Civil Engineering". Franck Schoefs is author of more than 35 peer reviewed journal papers, more than 123 international conferences and author of 6 chapters in books and co-editor of one book. He was awarded in 2011 the French Order of the Academic Palms as Chevalier.

IFSTTAR

IFSTTAR

IFSTTAR, the French Institute of Science and Technology for Transport, Development and Networks born on January 1st 2011 from the merger of INRETS and LCPC, is a major player in the European research on the city and the territories, transportation and civil engineering.

In order to be able to cope with the changing needs of society, IFSTTAR carries out researches and develops new technologies in the areas of urban engineering, civil engineering, infrastructure, construction materials, natural hazards and transportation of persons and goods. IFSTTAR investigates the uses and impacts of innovative solutions and decision-making tools from the technical, economic, social, health, energy, environmental and human points of view.

Key figures: 1150 Staff members, 6 sites in France, 90 European projects, more than 50 exceptional facilities

www.ifsttar.fr

informatics mathematics

Inria

Public science and technology institution established in 1967, Inria is is the only public research body fully dedicated to computational sciences. Combining computer sciences with mathematics, Inria's 3,449 researchers strive to invent the digital technologies of the future.

Educated at leading international universities, Inria's researchers creatively integrate basic research with applied research and dedicate themselves to solving real problems, collaborating with the main players in public and private research in France and abroad and transferring the fruits of their work to innovative companies.

They published over 4,500 articles in 2013. They are behind over 270 active patents and 110 start-ups. In 2013, Inria's budget came to 235 million euros, 25% of which represented its own resources.

www.inria.fr

UNIVERSITÉ DE NANTES

Université de Nantes

Université de Nantes is a major higher education and research centre in Western France. It is one of the rare French universities to promote interdisciplinarity. Within 21 faculties and schools, 295 specialities are represented and 63 laboratories work in all fields of knowledge. 75 % of them are A and A+ rated.

In a context of social and economic change, the capacity to innovate is more than ever the keystone of any competitiveness. Université de Nantes actively contributes to the innovation of the socioeconomic make-up through its researchers, its laboratories, its equipment and state-of-the-art technological platforms, its recent graduates who constitute a tremendous pool of skills and talent.

Key figures: 34,000 students (3,750 of them come from abroad) and 4,500 staff - 50% of them are dedicated to research.

www.univ-nantes.fr

Head of Local Organizing Committee

V. Le Cam, IFSTTAR, France
L. Mevel, Inria, France
F. Schoefs, Université de Nantes, France

Conference Secretariat

E. Blin, Inria, France
E. Lebret, Inria, France

Members of Local Organizing Committee

O. Abraham, IFSTTAR, France
C. Boisrobert, Université de Nantes, France
F. Bourquin, IFSTTAR, France
J.M. Caussignac, IFSTTAR, France
X. Chapeleau, IFSTTAR, France
M. Döhler, Inria, France
E. Gantier, IFSTTAR, France
C. Kouzoubachian, PRECEND, France
F. Lanata, École Supérieure du Bois, France
M. Le Pen, IFSTTAR, France
Y. Lecieux, Université de Nantes, France
D. Leduc, Université de Nantes, France

Contact

contact@ewshm2014.com

G. Akhras, Royal Military College, Canada

D. Balageas, ONERA, France

C. Boller, Saarland University, Germany

H. Budelmann, University of Braunschweig, Germany

M. Buderath, EADS, Germany

F. Casciati, University of Pavia, Italy

P. Cawley, Imperial College, UK

F.-K. Chang, Stanford University, USA

A. Cusano, University of Sannio, Italy

G. De Roeck, Catholic University Leuven, Belgium

S. D. Fassois, University of Patras, Greece

P. Ferdinand, CEA, France

C.-P. Fritzen, University of Siegen, Germany

M. Giordano, University of Naples, Italy

V. Giurgiutiu, University of South Carolina, USA

A. Güemes, UPM, Spain

W. Habel, BAM Berlin, Germany

J. Holnicki-Szulc, IFTR Warsaw, Poland

D. Inaudi, SMARTEC, Switzerland

G. Kawiecki, Boeing Research & Technology Europe, Spain

J. Kullaa, Helsinki Metropolia University of Applied Sciences, Finland

T. Kundu, University of Arizona, USA

V. Le Cam, IFSTTAR, France

L. Lecce, University of Naples, Italy

N. Maia, Technical University of Lisbon, Portugal

A. Mita, Keio University, Japan

Y.Q. Ni, Honk Kong Polytechnic University, China

W. Ostachowicz, Polish Academy of Sciences, Poland

C. Paget, Airbus Operations, UK

C. Papadimitriou, University of Thessaly, Greece

U. Peil, Technical University Braunschweig, Germany

S. G. Pierce, University of Strathclyde, UK

J. Rodellar, Technical University of Catalonia, Spain

M. Ruzicka, Technical University Prague, Czech Republic

M. Salvia, Ecole Centrale Lyon, France

D. Saravanos, University of Patras, Greece

D. Stargel, US Air Force, Columbia, USA

W. J. Staszewski, AGH University of Science and Technology, Poland

T. Stepinski, Uppsala University, Sweden

Z. Su, Honk Kong Polytechnic University, China

N. Takeda, University of Tokyo, Japan

M. Tur, Tel-Aviv University, Isreal

T. Uhl, AGH University of Science and Technology, Poland

SCIENTIFIC REVIEW COMMITTEE AND CHAIRMAN

For their participation to EWSHM 2014 as chairman and/or reviewer and/or keynote speaker, the organizing committee would like to thank:

O. Abraham
G. Akhras
D. Alleyne
P. Argoul
D. Balageas
D. Bernal
M. Billeres
C. Bockenheimer
L. Bodelot
C. Boller
L. Bouillaut
F. Bourquin
J. Brouns
P. Casari
F. Casciati
J-M. Caussignac
P. Cawley
F-K. Chang
X. Chapeleau
E. Chatzi
G. Cumunel
A. Cusano
X. Dérobert
P. Dersin
M. Döhler
M. Drissi-Habti
J. Dumoulin
L. Faravelli
S. D. Fassois
P. Ferdinand
V. Feuillet
E. Flynn
M. François
C-P. Fritzen
M. Frövel
M. Giordano
H. Gigleux
V. Giurgiutiu
B. Glisic
C. Gogu
P. Gueguen
A. Güemes
W. R. Habel
F. Hille
P. Hornych
F. Hoffmann
J. Holnicki-Szulc
B. Jacob
F. Jacquemin
A. Jeromin
Y. Kaya
F. Kopsaftopoulos
P. Kraemer
J. Kullaa
F. Lanata
B. Lebental
V. Le Cam
L. Lecce
Y. Lecieux
D. Leduc
A. Le Duff
E. Le Gal La Salle
B. L'Henoret
L. Mevel
F. Michelis
A. Mita
A. Nassiopoulos
Y-Q. Ni
E. Niederleithinger
P. Omenzetter
A. Orcesi
W. Ostachowicz
P. C. Ostiguy
C. Paget
V. Pakrashi
A. Paleologue
C. Papadimitriou
Y. Petryna
S. G. Pierce
P. Pons
E. Ramasso
J. Rodellar
M. Ruzicka
J. Sakellariou
J. Salin
N. Salowitz
M. Salvia
A-B. Samé
D. Saravanos
F. Schoefs
J-M. Simonin
F. Soldovieri
D. Stargel
W. J. Staszewski
P. Stephan
D. Stargel
T. Stepinski
Z. Su
D. Tcherniak
L-D. Theroux
M. Todd
V. Tournat
F. Treyssède
M. Tur
T. Uhl
C. Ventura
G. Villain
J. Waeytens
W-H. Wu
D. Zonta

Exhibition

1. Advitam
2. POLYTEC
3. EOTECH SA
4. FiberSensing
5. Université de Nantes
6. Mistras
7. IXFiber
8. OSMOS
9. Cofrend
10. A3IP
11. SCAIME
12. Structural Vibration Solutions A/S
13. SITES
14. Kinemetrics, Open Systems & Services
15. Smart Fibres / Prescamex
16. 19th WCNDT 2016 (DGZfP)
17. AMS Technologies
18. National Instruments
19. 4DSP
20. PRECEND

19th WCNDT 2016 (DGZfP)

19th WCNDT 2016
World Conference on Non-Destructive Testing
June 13 – 17 in Munich Germany

Stand No. 16

CONTACT

Address: Max-Planck-Str. 6 - 12489 Berlin - Germany

Tel: +49 30 67807-0

E-mail: mail@dgzfp.de

Website: www.dgzfp.de/Startseite/About-DGZfP

PRESENTATION

The 19th World Conference on Non-Destructive Testing in 2016 (19th WCNDT 2016) in Munich, Germany is organised by the German Society for Non-Destructive Testing (DGZfP).

We invite all researchers, users, manufactures and service providers from all fields of NDT to make an active contribution to the organization of an interesting and informative programme by submitting an abstract for oral or poster presentation. The call for papers will be published in January 2015.

During the exhibition NDT service providers, NDT equipment developers and NDT research institutes will have the opportunity to present their newest developments, NDT applications and research results. Online booking for exhibition is available.

More: www.wncdt2016.com

4DSP

Stand No. 19

CONTACT

Address: 1210 San Antonio Street - Suite 801 - Austin TX 78701 - USA

Tel: +1 800-816-1751

E-mail: salesfos@4dsp.com

Website: www.4dsp.com

PRESENTATION

4DSP's fiber optic strain, temperature and 3D shape sensing technology represents a quantum leap forward for industries as diverse as automotive, aerospace, oil & gas, medical devices, and wind energy. It is changing the way engineers look at materials and the way they design their products. Far ahead of the curve, we ensure that our core values of integrity and product excellence translate into our customers' success. The future of fiber optics looks bright, and we are delighted to be the pace-setters.

4DSP's Services for Fiber Optics Sensing are technical services made available to customers or prospective customers interested in engaging 4DSP for the purpose of advanced onsite training, product integration, feasibility studies, technical research or any other request for the services of the 4DSP Engineering Department:

- Feasibility study
- Onsite training
- Fiber embedment and installation
- Results analysis
- Software development for advanced display of data
- Product integration

A3IP

Stand No. 20

CONTACT

Address: bâtiment Placel - Route de Vannes
44880 Nantes - France

Tel: +33 (0)2 40 94 78 41

E-mail: benoit.mathonnet@a3ip.com

Website: www.pegase-instrumentation.com

PRESENTATION

A3IP is a French company specialized in electronics and embedded systems (sensor networks, smart systems...). Its main product line is "PEGASE-instrumentation", which offers a wide range of products to monitor any structure:

- "Strain gauge": displacement gauges, temperature, accelerometer
- "CASC": acoustic supervision of bridge cables (rupture detection)
- "FISSUROMETER": follow-up of cracks propagation
- "MOTORWAYS / RAILWAYS": civil engineering and metallic structures monitoring
- "GPS": structural movement using a high-precision GPS (1cm)
- "POLLUTION": multi-parameters physiochemical measurements of urban pollution

Customers can also purchase generic development cards for their own needs, or create a full design from needs for an industrial product. We provide courses to learn about PEGASE-Instrumentation and develop own solutions. Key points: "UTC" microsecond precision in a wireless world, remote control in real time of several PEGASE-Instrumentation, wireless sensor networks, energy efficient, can be deployed everywhere powered by a solar panel, web-clouding monitoring interface, available on PC, tablet and smartphone. PEGASE-Instrumentation benefits from competences, expertise and feedback from the IFSTTAR lab, partners as CEREMA and Power-Lan and customers as EUROVIA, COFIROUTE, COLAS, SNCF, GE Energy, DIR and more.

Advitam

Stand No. 1

CONTACT

Address: 1 bis rue du Petit Clamart - BP 80018
78146 Vélizy cedex - France

Tel: +33 (0)1 46 01 85 00

E-mail: anne-laure.pappalardo@advitam-group.com
pierre-antoine.laeuffer@advitam-group.com

Website: www.advitam-group.com

PRESENTATION

Advitam is a group of Civil Engineers, Structural Engineers, electronic and software engineers who operate in the field of Infrastructure Management Solutions providers. We develop systems and applications that provide readily available, timely and accurate information for managers faced with the tasks of operation, maintenance, and rehabilitation of infrastructure. Advitam devotes itself to understanding the risks of deterioration of infrastructure and implements software solutions, inspection, maintenance and monitoring systems in order to ensure their durability and serviceability at lower cost.

Advitam assists public and private contracting authorities - state and local authorities, manufacturers, and concession holders - with managing their built heritage.

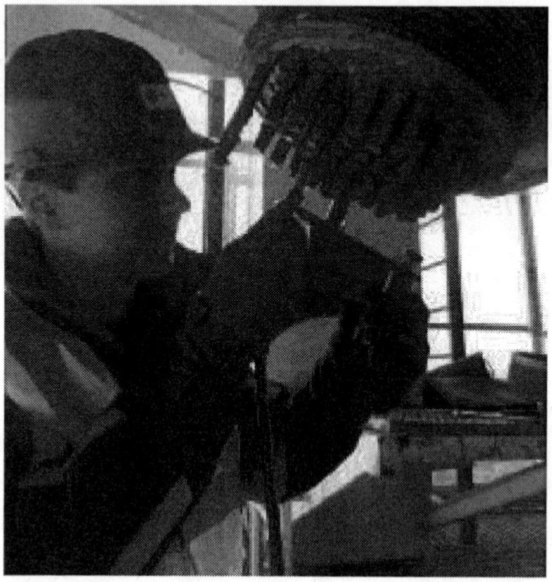

This is accomplished through 4 business areas:

- Infrastructure Management Software
- Remote Monitoring & Control Systems
- Investigations & Non Destructive Testing
- Cathodic Protection Systems

AMS Technologies

amsTECHNOLOGIES
where technologies meet solutions

Stand No. 17

CONTACT

Address: Fraunhoferstrasse 22 - 82152 Planegg
Germany

Tel: +49 (0)89 895 77 0 (Germany)
or +33 (0)1 64 86 46 00 (France)

E-mail: info@amstechnologies.com (Mr. Claudius RAITH)

Website: www.amstechnologies.com

PRESENTATION

AMS Technologies is Europe's leading solution provider and distributor for Optical, Power and Thermal Management Technologies, with 30 years of experience to date and currently serving more than 1000 European customers.

We are the specialists in both componentry and complete solutions for Optical technology, Thermal Management and Power Technology fields, with access to and long standing relationships with the most advanced manufacturers in each of those fields. Drawing extensively on our experience in each of these differing technologies, and coupling this with our broad system-level competence, we are able to offer seamless and comprehensive solutions incorporating complementary aspects from all three key technology fields.

Cofrend

COFREND
Confédération Française pour les Essais Non Destructifs

Stand No. 9

CONTACT

Address: 1 rue Gaston Boissier - 75724 Paris Cedex 15 - France

Tel: +33 (0)1 44 19 76 18

E-mail: pole.communication@cofrend.com

Website: www.cofrend.com

PRESENTATION

The French Confederation for Non-Destructive Testing is an organization open to any legal entity or any person concerned with non-destructive testing.

- For 50 years at the service of professionals in NDT, COFREND
  - Certifies the competence of Non Destructive Testing (NDT or NDT) operators
  - Is COFRAC accredited No. 0007 in accordance with EN ISO / CEI 17024
  - Conforms to EN ISO 9712, EN 410 4719/NAS, ISO 20807
  - Is Authorized ESP Pressure Equipment Directive 97/23 EC
  - Reckons more than 23,000 certified operators
- Animates, promotes and represents the French NDE network through its Governance and its four areas of expertise:
  - Professional Organization
  - Certification and Qualification
  - Science and Technique
  - Communication and Event

EOTECH SA

Stand No. 3

CONTACT

Address: 1 ZI du fond des Prés - 91460 Marcoussis France

Tel: +33 (0)1 64 49 71 30

E-mail: jj.servant@eotech.fr

Website: www.eotech-sa.com

PRESENTATION

Eotech proposes a wide range of products for measuring roughness, surface, forms, shape, geometries, dimensions, deformation, displacement, vibration, color, lighting and light signals for improving processes, control or inspection.

- High resolution 3D scanner
- Hand held 3D scanner
- Photogrammetry system
- Mobil CMM
- Roughness and profile measuring system
- 3D optical profiler

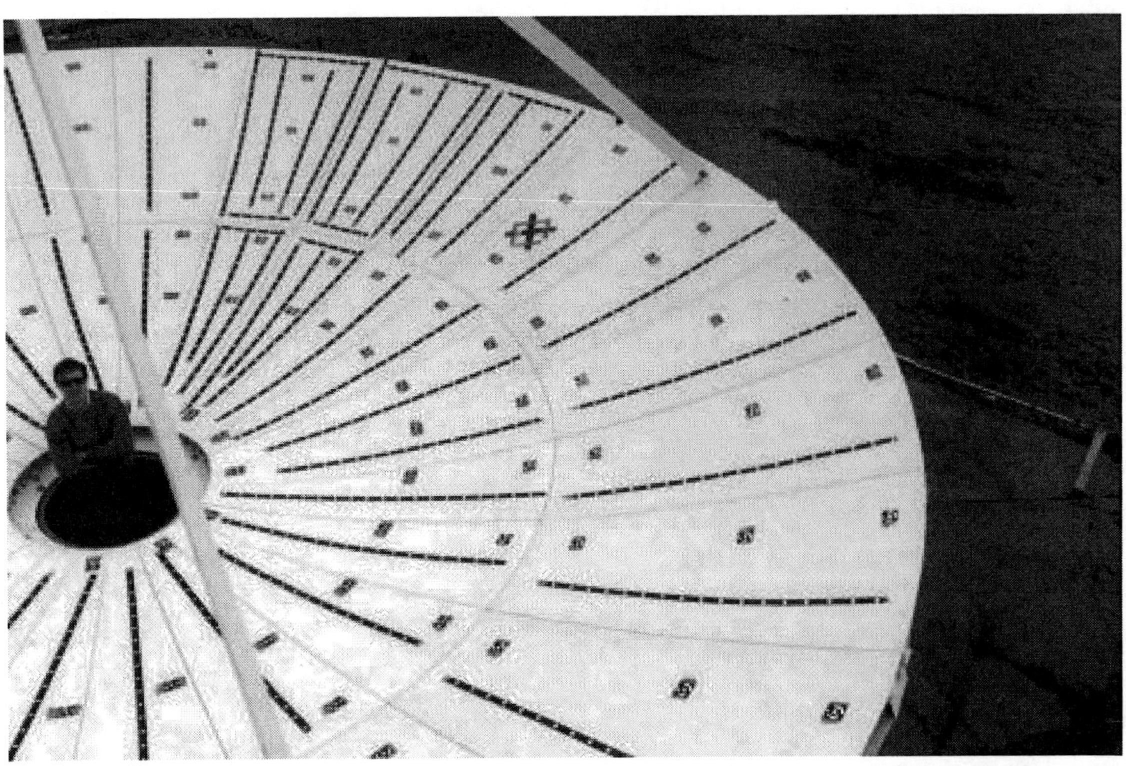

FiberSensing

Stand No. 4

CONTACT

Address: Rua Vasconcelos Costa - 277 - Maia
Portugal

Tel: +351 229 613 010

E-mail: info@fibersensing.com

Website: www.fibersensing.com

PRESENTATION

FiberSensing - Sistemas Avançados de Monitorizaçao, S.A. develops and produces in-house fiber optic sensor systems for advanced monitoring applications, based on Fiber Bragg Grating (FBG) technology. The company provides Sensors, Measurement Units, Software and Complete Solutions, primarily in an OEM basis, for the Construction, Energy, Aerospace, Industry and Transportation markets. Moreover, it has the ability to build customized monitoring solutions according to customers' specific requirements.

As a world leader in its field of expertise, FiberSensing is proud of having addressed more than one hundred different monitoring project around the world and delivered thousands of sensors and measurements units.

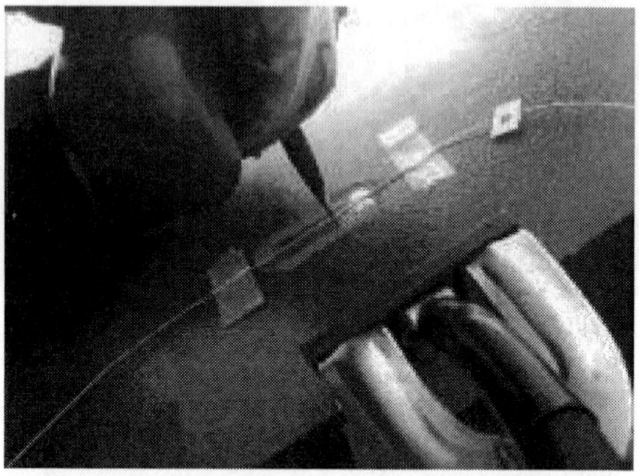

Monitoring solutions based on FBG technology have proven cost effective when applied to large scale structures where hundreds of sensors are deployed for long term measurement of different physical parameters. The use of FBG allows the drastic reduction of cabling due to the intrinsic high multiplexing capability of the technology, ensuring low impact on the instrumented structures.

The small size and weight of FBG sensors also make them particularly attractive for space restriction locations and embedding applications, such as in composite structures often used in bridges, wind blades, aircrafts, satellites, racing boat, among others. Actually, Fiber Bragg grating-based sensor systems have a huge application potential in many different sectors, such as Composites Industry, Laboratories, Research Centers and Universities looking for reliable measurements systems.

FiberSensing operates by recruiting Business Partners all over the world, which reinforce its presence in all five continents. Presently, the company has 39 collaborators working in 800 square meter state-of-the-art facilities located very near Porto International Airport, in Portugal. The company is certified in accordance with ISO 9001:2008 standards.

IXFiber

Stand No. 7

CONTACT

Address: Rue Paul Sabatier - 22300 Lannion - France

Tel: +33 (0)2.96.40.10.50

E-mail: patrice.crochet@xifber.com

Website: www.ixfiber.com

PRESENTATION

iXFiber is a leading independent manufacturer of active and passive specialty optical fibers, components based on Fiber Bragg Gratings (FBG) technology, and sub-assemblies modules for a diverse range of industries : Telecom, Sensing, Defense, Space, Fiber Laser, Harsh Environment...

- Innovative products in term of Specialty Optical Fibers and Fiber Bragg Gratings.
- Capability to design, develop and manufacture custom products.
- Capability to develop specific products using both technologies: Fiber and components.
- Space qualified Product. Harsh environment (radiation) qualified products.

Kinemetrics, Open Systems & Services

Stand No. 14

CONTACT

Address: 222 Vista Ave. - Pasadena - CA 91107, USA

Tel: +1-626-795-2220

E-mail: oss@kmi.com

Website: www.kmioss.com

PRESENTATION

Kinemetrics has been the world leader in the innovative design, quality manufacturing and timely supply of earthquake instruments for over 44 years. ISO 9001:2008 certified, Kinemetrics provides seismologists and structural engineers with the highest-quality, most cost-effective seismic instruments on the market. Open Systems & Services (OSS) goes beyond instrumentation; we are a world-leading, multi-disciplinary team of geoscientists, earthquake engineers, and information technologists. We specialize in the design, integration and implementation of custom seismic and structural monitoring systems. Our complete array of comprehensive systems and services, including project consulting and documentation, are designed to protect human lives, capital assets, and critical infrastructure investments around the globe.

Mistras

Stand No. 6

CONTACT

Address: 27 rue Magellan - 94370 Sucy en Brie - France

Tel: +33 (0)1 49 82 60 40

E-mail: contact@mistrasgroup.eu

Website: www.mistrasgroup.eu

PRESENTATION

MISTRAS is a leading "one source" global provider of technology enabled asset protection solutions used to evaluate the structural integrity of critical energy, industrial and public infrastructure.

With more than 40 years of infrastructure and structural experience, MISTRAS has become a name that clients can trust. As the largest and most experienced non-destructive testing (NDT) and structural health monitoring (SHM) company in the world, MISTRAS'expertise can be applied to many industries and clients: public infrastructures (bridges, ...) wind energy, nuclear industry, aerospace industry, refineries, power plants. Mission critical services and solutions are delivered globally and provide customers the ability to extend the useful life of their assets, improve productivity & profitability, comply with government safety and environmental regulations and enhance risk management operational decisions. ...

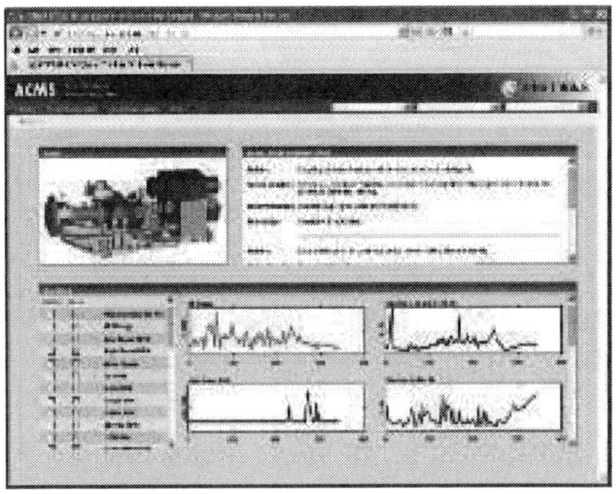

As the recognized leader of acoustic emission (AE) technology, MISTRAS designs, manufactures and markets AE sensors, systems and software for laboratory and on-line industrial clients. MISTRAS offers high-end ultrasonic imaging equipment and systems to a wide array of industrial applications, offers a variety of PdM services and vibration sensors, instruments and on-line plant asset integrity management systems for internet/intranet based predictive maintenance, and provides a complete range of inspection, training, QA/QC and proprietary Plant Condition Monitoring Software (PCMS).

With worldwide locations and over 5000 skilled employees, MISTRAS Group is the best choice for honest and reliable Structural Health Monitoring infrastructure needs.

National Instruments

Stand No. 18

CONTACT

Address: 2 rue Hennape - 92735 Nanterre cedex
France

Tel: + 33 (0)1 57 66 24 24

E-mail: ni.france@ni.com

Website: www.ni.com

PRESENTATION

For over 15 years, National Instruments has provided tools and solutions for asset condition monitoring applications spanning power generation, oil & gas, mining, transportation & heavy equipment, and manufacturing. Our extensive offering of products for machine diagnostics, on-line predictive maintenance and test cell applications includes embedded intelligent monitoring, computer plug-in monitoring boards, and software for vibration analysis and prognostics. These offerings are available in both an interactive configuration-based environment, NI InsightCM Enterprise, and the NI LabVIEW graphical programming environment.

National Instruments embedded control and monitoring systems, including NI LabVIEW software and NI reconfigurable I/O (RIO) hardware, combine the ruggedness and ease of use off-the-shelf systems like programmable logic controllers (PLCs), PC-based systems, and single-board computers (SBCs) with the specialization of custom hardware. All NI RIO hardware products (CompactRIO, Single-Board RIO) are built on an architecture that features powerful floating-point processors, reconfigurable FPGAs, and modular I/O. And with LabVIEW, the ability to customize hardware and integrate custom timing signal processing, and high-speed control without requiring expertise in low-level hardware description languages or board-level design.

OSMOS

Stand No. 8

CONTACT

Address: 5 rue Alfred de Vigny - 75008 Paris - France

Tel: +33 (0)1 71 39 95 15

E-mail: contact@osmos-group.com

Website: www.osmos-group.com/en

PRESENTATION

Founded in 2001, OSMOS is the world-wide leader in "structural health surveillance". Its cutting-edge technologies are used in 24 countries and have been established in Canada, Japan and Germany, through leading corporations. Its unique concept is based on the Optical Strand, a patented solution that is able to detect, in real time, structural weaknesses in all types of buildings. OSMOS offers more than just an innovative technology, it offers a true turnkey solution, with the only method capable of ultra-anticipatory flaw exposure on the state of any structure.

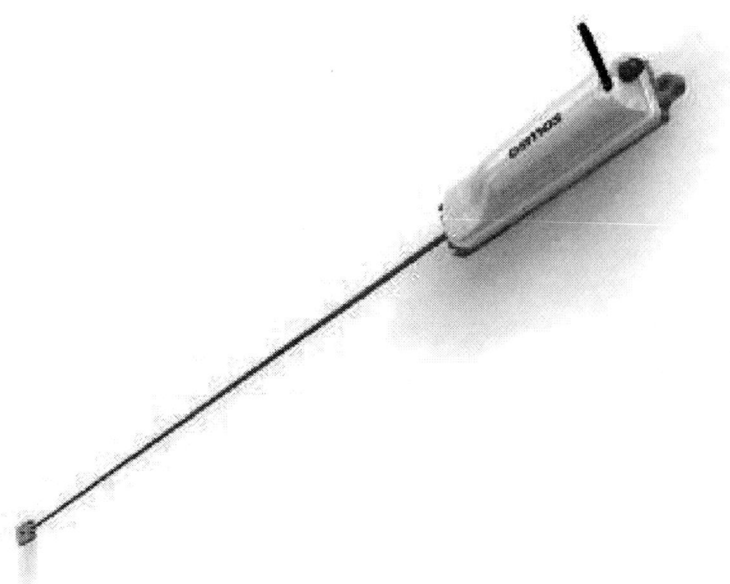

POLYTEC

Stand No. 2

CONTACT

Address: 39 rue Louveau - 92321 Chatillon cedex - France

Tel: +33 (0)1 49 65 69 00

E-mail: info@polytec.fr

Website: www.polytec.fr

PRESENTATION

POLYTEC Group, the market leader in electro-optical metrology since 1967, has nearly 400 employees worldwide and a turnover of € 60 million.

Manufacturer and distributor, Polytec offers optical solutions for fast and accurate measurement and control without contact, real-time acquisition, nondestructive testing and a wide range of complementary optical products.

The Polytec innovative solutions allow customers to maintain their own technical leadership across many fields in Public Domain (Education / Research) and Industry (Automotive, Aerospace, Defense, Electronics, Food...).

The French subsidiary of Polytec France SAS is organized around four technical and scientific activities among: laser vibrometry and profilometry, analytics, vision / photonics and polymers.

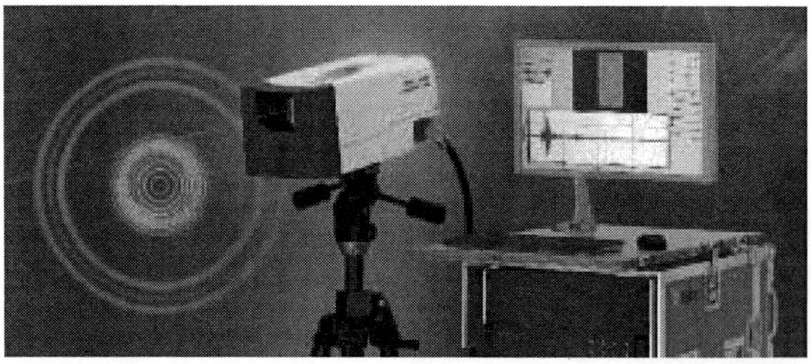

Precend

Stand No. 20

CONTACT

Cyril Kouzoubachian

Address: Precend c/o Capacités SAS - 26 boulevard Vincent Gache - 44200 Nantes - France

Tel: +33 (0)2 40 68 32 29 or +33 (0)7 77 20 26 70

E-mail: cyril.kouzoubachian@precend.fr

Website: www.precend.fr

PRESENTATION

Precend is a regional innovation platform which regroups forty centers of expertise in non-destructive testing (NDT) and structural health monitoring (SHM). Precend's objectives are animate, develop this network of expertise and make it available to the industry.

During the writing of this text, the following companies have already confirmed their participation on the shared booth Precend:

- **Alticontrole:** Services company with certified technicians according to EN 473 and ISO 9712: penetrant testing, magnetoscopy, ultrasound, TOFD, Phased Array, in accordance with your specifications and with applicable codes and standards.

More: www.alticontrole.com

- **ECND-PdL:** It is a scientific interest group that brings together the twenty laboratories present in Les Pays de la Loire region on the topics of NDT and SHM.

- **Image ET:** IMAGE ET offers solutions for externalized non-destructive X-ray control, with new technologies and open-source software developments. We are able to provide you an internal non-destructive control solution.

More: www.image-et.fr

- **Sciensoria:** Sciensoria sarl is specialist in model-based eddy current measurement of electrical conductivity and thickness. The measurement can be performed on both flat and curved surfaces, thin metal sheets or deposits, low conductivity materials like carbon composite (CFRP) and graphite, ferrous and other magnetic metals. Applications possible in aerospace, nuclear, pipeline and tank inspection, electrical machines and actuators design...

More: www.dtktor.com

- **Terahertz Waves Technologies:** Terahertz Waves Technologies designs and develops very innovative systems for measurement, inspection and characterization into the matter, providing 3D, real time, macro-molecular scale data by a non-destructive, non-invasive, contactless and safe technology.

More: www.t-waves-technologies.com

SCAIME

Stand No. 11

CONTACT

Address: 294, rue Gerorges Charpak - CS 50501
F74105 Annemasse Cedex - France

Tel: +33 (0)4 50 78 78 64

E-mail: info@scaime.com

Website: www.scaime.com

PRESENTATION

A team of specialists at your service...

Set up in 1983 with the added strength of a recognized technical heritage, SCAIME is one of the world leaders in the field of load cells and measurement for industry. With this full product range and its many associated services, SCAIME provides solutions enabling its customers' measuring problems to be solved efficiently with innovative solutions.

A global player with local presence

We are present in more than 60 countries through a network of subsidiaries and distributors. Our head office, located in Annemasse, houses our Sales & marketing services, the R&D department and assembly. We also benefit from group production means based in Shenzhen (China). This integrated, flexible and modern set of production solutions guarantees the best quality-price ratio for unit deliveries or for large series.

A philosophy: innovation and quality

All our activities are ISO 9001:2008 certified and we guarantee that our products are conforming to international standards and codes. This commitment to the continued improvement of our processes is intended to create a sustainable long-term relationship with our customers.

SCAIME, the best solutions to meet your requirements

Combining a renowned product quality and technological expertise, we are able to offer a complete range of measurement solutions based either on optical fiber or on traditional technologies and dedicated to industrial applications:

- Load cells
- Force sensors
- Extensometers
- Torquemeters
- Pressure sensors
- Displacement sensors
- Level sensors
- Fiber optics measurement
- Electronics

MDX400T-X Industrial Bragg sensors acquisition unit

Robust and reliable, MDX400T-X Bragg grating sensors conditioner provides high performance by conditioning up to 64 sensors on four optical lines at 100Hz.

It is specifically designed for installation in harsh environments and comes in an IP66 stainless steel housing that has passed certifications to sustain to high levels of shock and vibration.

MDX400T-X can work independently with internal memory for data storage. It embeds a Web server for remote setup and Etherlink for data transfer. When connected to URi 3G router it can be remotely operated (setup, sms and email alarms, data and files transfer, restarts...).

SITES

Stand No. 13

CONTACT

Address: 2 bis avenue du centre
92500 Rueil-Malmaison - France

Tel: +33 (0)1 41 39 02 00

E-mail: contact@sites.fr

Website: www.sites.fr

PRESENTATION

Recognized specialist in studies, inspection, controls and preventive monitoring of ageing structures and environments, SITES quickly imposed itself in the world of nuclear power plant, bridge, dam manager, infrastructures... The company, since its creation in 1984, uses, develops, implements, combines the skills and performance tools to make sure of their durability and smooth operation of these structures.

Our success in France (5 offices - Paris, Lille, Tours, Lyon, Aix-en-Provence) has allowed us to export our know-how. This opening up to globalization is characterized by the creation of permanent offices in China (SITAI) and South Africa (SITES AFLA).

EDF, Bouygues TP, Areva, Aéroport de Paris, SNCF, Vinci, Le Louvre... trust us and work with us since years.

Smart Fibres

Stand No. 15

CONTACT

Address: Brants Bridge - RG12 9BG Bracknell - UK

Tel: +44 (0) 1344 484111

E-mail: info@smartfibres.com

Website: www.smartfibres.com

PRESENTATION

Smart Fibres pioneered the development of fiber Bragg grating based monitoring systems and has been an industry leader for over 15 years.

With a broad range of FBG fiber optic sensor and interrogator products and a wealth of applications knowledge, we serve the world's leading companies and institutes by developing practical fiber optic sensing solutions in many market areas including oil and gas, aerospace, renewable energy, marine, and civil engineering.

As an ISO-9001 accredited company, Smart Fibres is highly committed to consistent quality manufacturing and efficient business operations. Our current developments are set to make fiber optic sensing more broadly available and economically viable as a performance optimization and asset management tool in a number of key volume sectors.

PRESCAMEX

Stand No. 15

CONTACT

Address: 10 Rue des Ardennes - 77123 Le Vaudoué - France

Tel: +33 (0)1 60 39 61 61

E-mail: contact@prescamex.fr

Website: http://prescamex.fr.pagesperso-orange.fr/presentation-en.html

PRESENTATION

PRESCAMEX, the leader of extensometry in France and Europe, offers highly recommended skills for the installation of sensors and systems, as well as data acquisitioning and stress analysis, but also in the choice of strain gauges, and building optical fiber FBG sensor systems.

Based in France, south of Paris, Toulouse, Nantes and Istres, PRESCAMEX measurements laboratory uses the latest up-to-date technologies to satisfy the needs of its clients.

Technicians and Engineers are trained to the latest technologies, data acquisition systems and products, to satisfy customers in more and more specific environmental conditions. Our team will work in the tightest lead time in France and abroad to install sensing systems on specimens, buildings, large engineering equipment, petrol off shore platform, aircraft...

PRESCAMEX, ISO9001 accredited company, is proud to be the exclusive distributor of all Smart Fibers sensors and interrogators for the French market.

Structural Vibration Solutions A/S

Stand No. 12

CONTACT

Address: NOVI Science Park - Niels Jernes Vej 10
9000 Aalborg - Denmark

Tel: +45 9635 4422

E-mail: sales@svibs.com

Website: www.svibs.com

PRESENTATION

Structural Vibration Solution A/S was founded in 1999 as a spin-off company from Aalborg University in Denmark. We are focused on the mission to be the leading provider of Operational Modal Analysis solutions in the world.

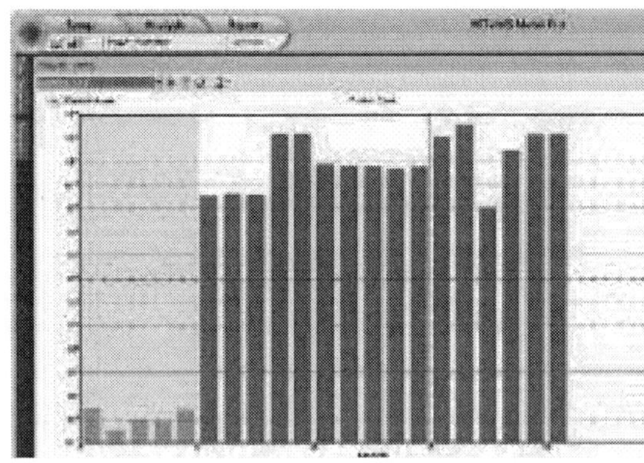

Our software ARTeMIS helps our customers to determine and solve vibration issues in the vast number of cases where the excitation cannot be measured or controlled. The software is used by engineers all over the world for modal analysis of all kinds of structures.

We are introducing a new Statistical Subspace-based Damage Detection Module in the upcoming ARTeMIS Modal Pro to be presented at the EWSHM 2014.

Université de Nantes

UNIVERSITÉ DE NANTES

Stand No. 5

CONTACT

Franck SCHOEFS

Address: Université de Nantes, Master of Civil Engineering - 2 rue de la Houssinière 44322 Nantes cedex 3 - France

Tel: +33 (0)2 51 12 55 22

E-mail: franck.schoefs@univ-nantes.fr

Website: www.univ-nantes.fr/sciences/mastergc

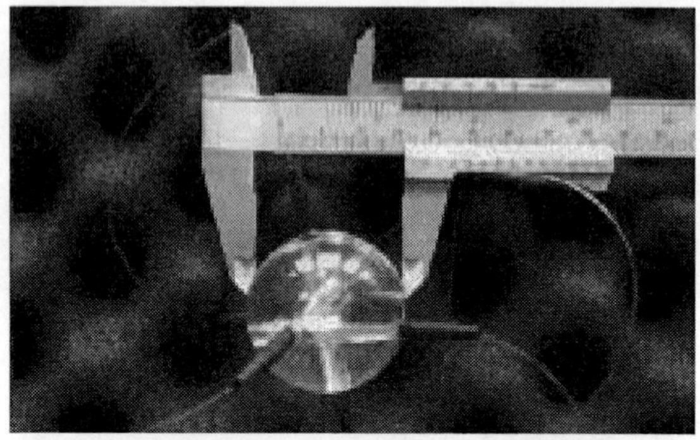

PRESENTATION

Masters of Sciences in Mechanical and Civil Engineering provide an expert knowledge in structural and material engineering, numerical methods and more specialized courses in monitoring technology, data statistical analysis and modeling, maintenance optimization, life cycle analysis and probabilistic modeling. Their team are involved in GeM laboratory (UMR CNRS 6183, Université de Nantes, Ecole Centrale de Nantes). When focusing on SHM, this team is engaged in various partnerships with big societies (TOTAL, EDF, STX, DCNS, Port Authorities, ...) and SMEs (OXAND, PHIMECA, ...) has two main patents and publications in peer-reviewed journals (more than two per academic position per year) and projects supported by national governments (French: MAREO, EVADEOS, ...) or European Community (MEDACHS, DURATINET I and II). Finally this group is involved in the commercial subsidiary of Université de Nantes, CAPACITES with a business unit called IXEAD. Université de Nantes will illustrate these competences in pedagogy and research through at least two illustrations during the exhibition: a physical model of a wharf monitored with Fiber Optical Sensors for pedagogical applications and a patent for measuring local 3-D strains in massive materials (concrete dams....).

General information

Registration, information and Tour desks

Registration, information and tour desks

Location:

All the EWSHM 2014 services desks
are located at the main entrance of the convention center's mezzanine.

Opening hours:

Monday 7th: 17h30-19h00,
Tuesday 8th: 7h45 - 19h00,
Wednesday 9th: 7h45-18h00,
Thursday 10th: 7h45-18h00,
Friday 11th: 7h45-13h00,

EWSHM 2014 staff will be available to offer you on-site assistance and advice.

Registration is still open at the registration desk until the end of the conference, but you can only pay by credit card. Registration includes a name badge, vouchers, the conference program & abstract book and a USB stick of the conference proceedings.

Name badge

Please wear your conference name badge at all times. Access to the conference location, social events and visits will not be granted without it.

Disabled access

The entire venue is wheelchair accessible. See and follow the wheelchair symbol sign in the convention center for the location of the lifts.

First Aid

In case of problems, please contact any member of the EWSHM 2014 staff; there is a team of trained first-aiders and cardiac first-responders on site at all times.

Food allergies

If you have any questions about the food served at the conference, please contact the staff at the buffet tables, they will be able to advise you regarding the food and any particular requirements you may have.

Smoking policies

The Nantes international events center is a non-smoking area. Smoking is only possible outside.

Personal property

The participants are invited to take good care of their personal belongings, and to not leave them unattended. Neither the organizers nor the staff will be responsible for any loss or damage of the personal property of the participants. Please contact the EWSHM 2014 information desk in any case if something is lost.

Lunch and coffee breaks

From Tuesday, July 8 to Friday, July 11, coffees and refreshments will be served on the mezzanine in the morning at 10:00 and in the afternoon at 16:00.
From Tuesday, July 8 to Thursday, July 10, lunches will be served in the area R2 at noon. For Friday, July 11, you can find in your conference bag a ticket for a lunch box.

Free wifi

Free wireless internet service (SSID: EWSHM) is accessible throughout the entire venue.

Presenting an oral paper (keynote or technical paper)

Oral presentations are organized by sessions, as indicated in the technical program. They are scheduled in specific lecture rooms together with the time of presentation of each contribution including discussion and change over. The time allocated for each keynote lecture is 30 minutes including 5 minutes for questions and discussion. The time allocated for each technical presentation is 20 minutes including 5 minutes for questions.

→

→

Authors are kindly asked to upload their presentations directly in the respective lecture room at least 20 minutes before the start of the session. A lecture room assistant will be available for any help. Each room has a laptop, a video projector and a microphone. The laptops provide PowerPoint, Acrobat Reader and VLC.
It is strictly prohibited to take photos and/or copies from notebooks of any scientific material without the expressed permission of the authors.

Presenting a poster
The poster boards are portrait having maximum dimensions of 841mm width by 1189 mm height. Posters should be put up in the Area R2 between 08:00 and 09:00 in the morning of Tuesday, July 8th, 2014. A student assistant will help you locate the panel for your poster.

Useful information about Nantes
A hostess of the tourist office of Nantes is present at the registration desk for information about Nantes. More information about the tourist office of Nantes at www.levoyageanantes.fr/en

A map of Nantes and a guide of public transportation are available in the conference bag.
More information about public transportation at www.tan.fr

The staff of EWSHM 2014 will wear a red T-shirt during the entire conference: please do not hesitate to contact them for information about the conference, transport and the beautiful city of Nantes and surrounding areas.

The Local Organizing Committee has been doing its best to offer you an unforgettable stay in Nantes during the conference. Original, traditional and cultural places of Nantes pave the ground of a promising social program. Besides all the lunches, we offer you the following social program:

Monday, 7 July 2014

Icebreaker reception

At La Cité, the Nantes Events Center
From 17h30 to 19h00
All attendees of EWSHM 2014 are invited to the icebreaker reception

Get to know each other in a moment of conviviality.

Offered by
EWSHM 2014

Tuesday, 8 July 2014

Buffet dinner

At La Cité, the Nantes Events Center
From 18h00 to 20h00
All attendees of EWSHM 2014 are invited to the buffet dinner

Take time to visit exhibitors as well as the poster session while tasting French foods and wines.

Offered by
Nantes Metropole

Wednesday, 9 July 2014

Gala dinner

At Les Machines de l'île de Nantes in the Elephant Hall
From 19h30 to 23h30
All registered people will find their invitation in their conference kit

Sponsored by
AIRBUS

Thursday, 10 July 2014

Boat cruise and dinner

Boarding for a boat cruise to go to Le Château de la Poterie
where dinner will be served
From 18h45 to 23h30
All registered people will find their invitation in their conference kit

Offered by
EWSHM 2014

Wednesday, 9 July 2014: **gala dinner**

Les Machines de l'île de Nantes (Shipyards Park)

The Elephant hall is one of the large halls of glass, concrete and steel created in the early 20th century in the south of the island of Nantes. They sheltered workshops boiler of Chantiers de la Loire for equipment of ships. After rehabilitation, the area now features a completely new artistic project "Machines de l'île". Born from the imaginations of François Delarozière and Pierre Orefice, it is at the crossroads of Jules Verne's "invented worlds", the mechanical universe of Leonardo da Vinci, and of Nantes' industrial history. Thus, some strange machines came to populate the Île de Nantes...

Location and shuttle information

Les machines de l'île de Nantes is located at Boulevard Léon Bureau (2,4 km from la Cité).

A complimentary shuttle service is provided for all participants for the return of the evening only. More information will be given during the conference.

Access: see map on the next page →

By walk: 30 minutes
By bus and tramway: less than 15 minutes
1/ Bus #4: station Cité Internationale des Congrès → station Duchesse Anne-Château (direction Foch-Cathédrale)
2/ Tramway #1: station Duchesse Anne-Château → station Chantiers Navals (direction Jamet / François Mitterrand)

Thursday, 10 July 2014: **boat cruise and dinner**

Le Château de la Poterie

Located in the heart of Val d'Erdre ("the most beautiful river in France" according to Francis 1st), this castle (manor) was built in the late eighteenth century on the ruins of an ancient roman tile. Fireplaces, old carved stone from Loire, old beams: the setting is exceptional with a view of the Erdre... the evening will take place in an outbuilding of the castle, in a warm and friendly atmosphere after a boat cruise on the Erdre river.

Location and shuttle information

The pier for the boat cruise is located at Quai de la Motte Rouge (1,9 km from la Cité.).

A complimentary shuttle service is provided for all participants for the return of the evening only. More information will be given during the conference.

Access: see map on the next page →

By walk: 25 minutes
By bus and walk: 15 minutes
1/ Bus #4: station Cité Internationale des Congrès → station Duchesse Foch-Cathédrale (direction Foch-Cathédrale)
2/ Walk: 12 minutes

Wednesday, 9 July 2014: gala dinner
Access to les Machines de l'île de Nantes

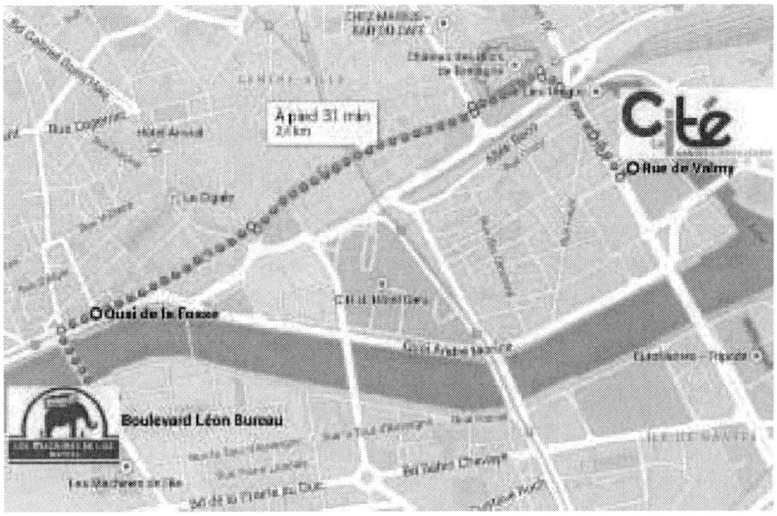

Thursday, 10 July 2014: boat cruise and dinner
Access to the pier for the cruise towards Le Château de la Poterie

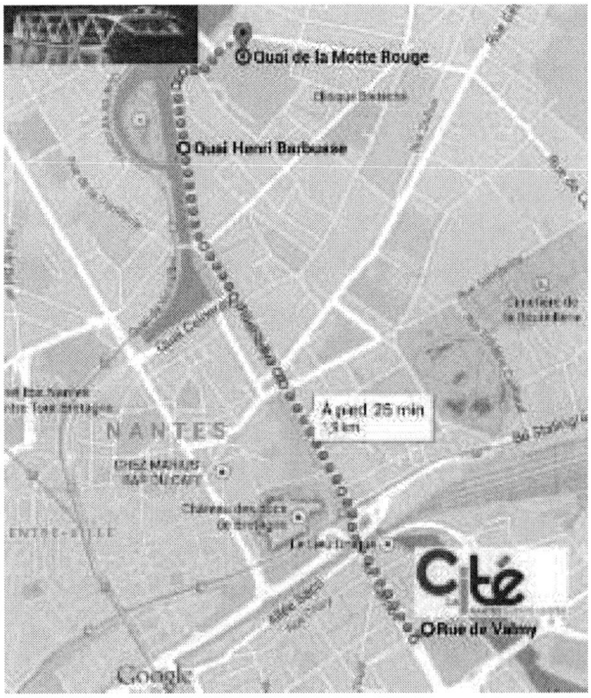

AIRBUS - Nantes

Specialized in the manufacturing of centre wing boxes for all Airbus aircrafts, Nantes has acquired advanced expertise in the use of composite materials and the machining of large-scale, complex, aluminium alloy parts.

AIRBUS - Saint Nazaire

Saint-Nazaire is the place where Airbus forward and centre fuselage sections are integrated. These sections are made up of several assemblies manufactured throughout the world. Once assembled, the sections are fitted with all the aircraft's major systems (hydraulic, fuel, electrical) and tested, before being sent to the final assembly lines. As a hub between the assembly lines in Toulouse, Hamburg, Seville and Tianjin (China), Saint-Nazaire delivers fuselage sections via cargo aircraft (Beluga) and, for the A380, an integrated "road-sea-river" transport link.

Note: Airbus Saint Nazaire is located at one hour by bus from Nantes.

IFSTTAR

One of the specific characteristics of the IFSTTAR site at Nantes is its large scientific equipments. During the visit, you can discover:

- Geotechnical centrifuge
- Accelerated load testing facility: the fatigue carousel
- Cable fatigue test bench for civil engineering structures
- Testing facilities and composite footbridge

IRT Jules Verne

The IRT Jules Verne is a mutualized industrial research institute dedicated to advanced manufacturing. At a single site in Nantes, it brings together manufacturers, training facilities, private and public applied research laboratories, prototyping and industrial demonstration resources. The visit concerns its technological platform "Technocampus Composites" that gathers 25 organizations working on developing innovative processing technologies for high-performance composites materials. During the visit you can discover some of the equipment used for R&D projects and pre-industrialization processes including: filament winding machinery, thermocompression and pultrusion line, robotic fibres placement, non-destructive testing...

SHM Person of the Year Award 2014

This award is sponsored by 'SAGE'

A structural health monitoring person of the year (SHM-POY) will be selected by the editors and associate editors of Structural Health Monitoring: An International Journal. The Person of the Year should have made an outstanding contribution to the field of SHM that will benefit society. This contribution can be in the form of theory, analysis, applications, education, or other ways that support the discipline of SHM and benefit society. The award is meant to recognize accomplishments within the past year or few years.

Achenbach Young Investigator Award

This award is sponsored by 'EWSHM 2014'

EWSHM 2014

The Achenbach Medal has been created to recognize an individual (within 10 years of PhD) who has made an outstanding contribution to the advancement of the field of Structural Health Monitoring. It is an international award and nominations from all over the world are encouraged. This award is will be announced by Pr. Fu-Kuo Chang (Stanford University).

Best Paper Award

This award is sponsored by 'EWSHM 2014'

EWSHM 2014

The SHM Best Paper Award is presented to one or more individuals whose paper(s) are selected to have the highest quality and innovation from the EWSHM 2014 proceedings. The Award Committee is formed by all the session chairmen for this workshop. They may propose among the articles from the sessions of their specialty, one or two papers which may be candidates for the Award, justifying their proposal. This award is sponsored by Inria, IFSTTAR and Université de Nantes.

Student Best Paper Award

This award is sponsored by 'ECND-PDL'

A research pole for Non Destructive Testing and Evaluation of la Région Pays de la Loire (France)

A student best paper award will be presented at the EWSHM 2014. Candidates need to be first author of the paper, present the paper themselves at the EWSHM 2014 and either have a valid student ID at the time of registration or have defended the thesis after 1 January 2013. Originality and technical excellence are the two main criteria for selection. The Award Committee is formed by members of ECND-PDL that sponsors the student best paper award of this workshop and invited foreign attendees. The contribution of Bernard Jacob (Scientific Direction, Ifsttar) to the Student Best Paper Award process is kindly acknowledged.

Fu-Kuo CHANG

Professor in the Aeronautics and Astronautics Department - Stanford University, USA

PRESENTATION

The need of SHM Quantification for Implementation

Tuesday, July 8th, 9:00-9:30 - Auditorium 450

BACKROUND

Dr. Fu-Kuo Chang is a Professor and the Director of Structures and Composites Laboratory in the Department of Aeronautics and Astronautics at Stanford University. He received his PhD. from University of Michigan in 1983. His specialties include structural health monitoring (SHM); multifunctional materials and structures; and damage tolerance and design of composite structures. He is the Editor-in-Chief of the international Journal of Structural Health Monitoring, the recipient of 2004 SHM Life-time Achievement Award by Boeing Company, and the 2010 SPIE Smart Structures/NDE Life-Time Achievement award; Fellows of AIAA and ASME.

Christian BOLLER

Chair in NDT & QA - University of Saarland, Saarbrücken, Germany

PRESENTATION

Structural Health Monitoring - What WE need to do now

Tuesday, July 8th, 9:30-10:00 - Auditorium 450

BACKROUND

Christian Boller holds a chair in Non-Destructive Testing (NDT) and Quality Assurance at Saarland University in Saarbrücken/Germany and is also a Director of the Fraunhofer Institute of Non-Destructive Testing based in Saarbrücken/Germany since 2008. He received an engineering diploma in structural engineering from Darmstadt Technical University as well as an engineering doctoral degree in material mechanics from the same institution. He held various posts in research and development with Battelle, MBB, Daimler-Benz and EADS in Frankfurt/Main, Ottobunn/Munich and Stuttgart respectively, before being appointed a chair in smart structural design at the University of Sheffield/UK in 2003. Professor Boller is a member of various scientific committees and societies and has published and edited a multitude of articles and books in the area of fatigue and fracture, smart structures, structural health monitoring (SHM) and micro aerial vehicles, including Encyclopedia on Structural Health Monitoring. He has become the major organiser of a variety of international conferences such as European Workshop on SHM, NDT in Aerospace or NDT for the Energy Industry to just name a few. His major research areas of interest include SHM inspired by NDT and integrated into structural life cycle management as well as robotic inspection systems. Since 2013 he is also running an NDT Master Course at Dresden International University and launched a module on SHM at Saarland University in 2014.

Falk HOFFMANN

Airbus Helicopter Deutschland GmbH, Germany

PRESENTATION

PHM: Enabling Condition Based Maintenance for Helicopters

Wednesday, July 9th, 8:30-9:00 - Auditorium 450

BACKROUND

Mr. Hoffmann is currently the Head of the Health and Usage Monitoring Systems (HUMS) Department for Airbus Helicopter in Germany. Prior to his present position he worked as a Development Engineer in the Dynamics and Vibration Department of Eurocopter. In addition, Mr. Hoffmann worked as a fellow researcher at Daimler Chrysler. Between 2000 and 2007, he was part of EADS Innovation Works. Mr. Hoffmann's primary research fields are in HUMS, Condition Based Maintenance (CBM) and active noise and vibration control. His academic publication includes 3 patents and over 16 international publications, making him an expert in the field of HUMS and CBM.

Pierre DERSIN

Alstom Transport Information Solutions, France

PRESENTATION

PHM: An attempt at diagnosing the present and prognosticating the future...

Wednesday, July 9th, 9:00-9:30 - Auditorium 450

BACKROUND

Dr. Dersin obtained his Ph.D. in Electrical Engineering in 1980 from the Massachusetts Institute of Technology (MIT) after receiving a Master's degree in O.R. in 1976, also from MIT. He worked on reliability of large electric power networks, as part of the Large Scale System Effectiveness Analysis Program sponsored by the US Department of Energy, from MIT and Systems Control Inc. and later joined FABRICOM (Belgium and U.S), where he was involved with fault diagnostic systems for factory automation, from 1983 to 1999. Since 1990 with ALSTOM Transport, he has occupied several positions, mainly involved with RAMS and Maintenance, including R&D Manager of the Service business since 1999, and, since 2007, RAM Director at ALSTOM Transport's Information Solutions. Also, he is a Leader of ALSTOM Transport's Reliability & Availability Core Competence Network, and since April 2014, a co-Director of the joint ALSTOM-INRIA Research Laboratory for digital technologies applied to mobility and energy. He has several publications in IEEE conferences and journals. He was the chairman of UNIFE's (Union of European Railways Manufacturers) LCC/RAM working group in 1998-2001. He is an associate editor of the International Journal of Performability Engineering, He is a member of the IEEE Reliability Society's Advisory Committee and IEEE Industry Advisory Board.

Claus-Peter FRITZEN

University of Siegen, Department of Mechanical Engineering, Germany

PRESENTATION

Monitoring of Wind Energy Plants - Challenges and State of the Art

Wednesday, July 9th, 9:30-10:00 - Auditorium 450

BACKROUND

Dr. Claus-Peter Fritzen is a Professor of Applied Mechanics in the Department of Mechanical Engineering and currently the vice-chairman of the Center of Sensor Systems at the University of Siegen. His research interests include the development of vibration-based and wave-based methods for Structural Health Monitoring and their application in different engineering fields, such as wind energy plants, aircraft and civil structures. He is chief editor of the Intl. Journal of Structural Monitoring and Maintenance (SMM) and associate editor of the Intl. Journal of Structural Health Monitoring (SHM). In 2010 Prof. Fritzen received the SHM Person of the Year Award. He has published many papers and is coauthor of various books on SHM. He regularly teaches advanced courses together with other colleagues working in this field.

Peter KRAEMER

Woelfel Beratende Ingenieure GmbH & Co. KG., Germany

PRESENTATION

Monitoring of Wind Energy Plants - Challenges and State of the Art

Wednesday, July 9th, 9:30-10:00 - Auditorium 450

BACKROUND

Dr. Peter Kraemer received his PhD from the Department of Mechanical Engineering of the University of Siegen in Germany for his work on "damage diagnosis approaches for structural health and condition monitoring of offshore wind energy plants". Presently he is working as a development engineer and designer of SHM-systems for wind energy plants with Woelfel Beratende Ingenieure GmbH & Co. KG in Hoechberg, Germany. His main activities concentrate on the development of algorithms for data and signal analysis, vibration-based online monitoring of wind energy plants and other systems such as railway vehicles, machine tools, etc

Benjamin L'HENORET

Project Manager - Innovative monitoring systems, SNCF Innovation & Research, France

PRESENTATION

Track condition monitoring at SNCF: an introduction to methods, tools and perspectives

Thursday, July 10th, 8:30-9:00 - Auditorium 450

BACKROUND

He has been working for ten years in automotive and railway industry for new sensors development. His research interests include smart infrastructure, WSN, autonomous sensors and monitoring data processing for maintenance decision support tools. His is an active member of the SNCF Synapse Scientific & Techniques expertise group in the field of mechatronics.

Hugues GIGLEUX

Head of Track Maintenance Engineering Innovative Projects Group - SNCF Infrastructure, France

PRESENTATION

Track condition monitoring at SNCF: an introduction to methods, tools and perspectives

Thursday, July 10th, 8:30-9:00 - Auditorium 450

BACKROUND

He has been working for three years in the French railway infrastructure company (SNCF INFRA) on the deployment of innovative tools to improve condition monitoring of the network. He is especially involved in projects using video monitoring, WSN and/or data processing. He is an active partner of many research and innovation activities inside SNCF INFRA.

Jean SALIN

Senior Research Engineer - EDF R&D, France

PRESENTATION

Improvement of Civil Works Health Management EDF's Global Strategy and major advances between 2008 and 2013

Thursday, July 10th, 9:00-9:30 - Auditorium 450

BACKROUND

Graduated Engineer in Materials Sciences in 1978, M. Salin joined Electricité de France in 1981 in the Operating Nuclear Division. During years he has been educated in NDT techniques by old mates, and involved in In Service Inspection of different metallic components as reactor vessel, turbines, heat exchanger... From 1998 to 2007 he acts as vice chairman of the EDF's Qualification Body for NDT, in accordance with the French regulation. Since 2008, he serves as senior research engineer for inspection and monitoring of civil engineering structures, and more recently, gets interest in the ISI of onshore wind turbine. A NDT career from metallic to concrete then composite components!

Pierre STEPHAN

Research Engineer - EDF R&D, France

PRESENTATION

Improvement of Civil Works Health Management EDF's Global Strategy and major advances between 2008 and 2013

Thursday, July 10th, 9:00-9:30 - Auditorium 450

BACKROUND

Dr Pierre STEPHAN is graduated in materials science and mechanics in 2004. After 5 years making expertise on metallic and composite components damaged by fatigue and wear at CETIM's labs (Technical Center of mechanical Industries), he joins EDF R&D as research engineer. His research topics interest now SHM of nuclear Civil Works (Cooling towers, reactors containment) and material (primary pumps for instance), hydraulic ones (Dams...) and photovoltaic power plants. That led him to work especially on monitoring methodologies, rebar's corrosion, and vibration analysis. He has been working for ten years in automotive and railway industry for new sensors development. His research interests include smart infrastructure, WSN, autonomous sensors and monitoring data processing for maintenance decision support tools. His is an active member of the SNCF Synapse Scientific & Techniques expertise group in the field of mechatronics.

Carlos VENTURA

Professor of Civil Engineering, Department of Civil Engineering
- University of British Columbia, Vancouver, Canada

PRESENTATION

Structural Health Monitoring of Civil Engineering Structures - Lessons Learned and Future Directions

Thursday, July 10th, 9:30-10:00 - Auditorium 450

BACKROUND

Carlos Ventura is a Civil Engineer with specializations in structural dynamics and earthquake engineering. He has been a faculty member of the Department of Civil Engineering at the University of British Columbia (UBC) in Canada since 1992. He is currently the Director of the Earthquake Engineering Research Facility (EERF) at UBC, and is the author of more than 450 papers and reports on earthquake engineering, structural dynamics and modal testing. He is a member of the Canadian Academy of Engineering and Fellow of Engineers Canada, also a member of several national and international professional societies, advisory committees and several building and bridge code committees. Dr. Ventura has conducted research about earthquakes for more than thirty years. His research work includes experimental studies in the field and in the laboratory of structural systems and components. His current research includes the development and implementation of performance-based design methods for seismic retrofit of low rise school buildings, novel techniques for regional estimation of damage to structures during earthquakes, and on structural health monitoring of building, bridges and dams. In addition to his academic activities, Dr. Ventura is a recognized international consultant on structural vibrations and safety of large Civil Engineering structures. His consulting practice includes projects with companies and government institutions in North America, Central America, South America, Asia and Europe.

Frédéric BOURQUIN

Head of the Components and Systems Department - IFSTTAR, France

PRESENTATION

SHM trends and opportunities for the civil engineering sector

Friday, July 11th, 8:00-8:30 - Auditorium 450

BACKROUND

Frédéric Bourquin graduated in 1984 at Ecole Polytechnique in France and in 1987 at Ecole Nationale des Ponts et Chaussées. He defended in 1991 a PhD thesis in applied mathematics and his habilitation to supervise PhD theses in 1995. He is the author of over 50 peer reviewed publications and several patents. His research topics include numerical methods for fluid-structure interaction, dynamics of complex structures, inverse heat transfer, active control of structures, sensor placement, energy efficiency assessment of buildings, simplified models for smart grids, inverse potable water networks modeling, scour monitoring and numerical simulation of nano-transducers. He supervised 23 PhD students, four of them in Italy. He heads the Components and Systems department (300 scientists) at IFSTTAR with emphasis on research and technology transfer.
Co-leader of the Focus Area *Networks* and of the *refine* initiative at ECTP, member of the Joint Task Force on infrastructures and of the European *For ever Open Road* executive committee, he participated in the board of the International Association for structural Control. He acts as an expert for ESF and has chaired three international conferences on control and organized two. He is a member of the European Laboratory Lagrange and has contributed to more than five European projects on monitoring and control (ASSET, SAMCO, SMART, CONVIB, SMARTER, ISTIMES) as well as domestic industry-driven projects on the monitoring of transport infrastructures, water supply networks and water quality, smart grids for smart energy management of buildings.

Bruno GODART

Deputy head of the Bridges and Structures Department - IFSTTAR, France

PRESENTATION

SHM trends and opportunities for the civil engineering sector

Friday, July 11th, 8:00-8:30 - Auditorium 450

BACKROUND

Bruno Godart, born 1956, received a CE diploma from the Ecole Nationale des Travaux Publics de l'Etat in France in 1975, and a Master of Science from Stanford University in 1979.
Having done his career in the Laboratoire Central des Ponts et Chaussées (LCPC - Central Laboratory for Roads and Bridges) where his last position was Technical Director for Bridges, Bruno GODART is currently the deputy head of the Bridges and Structures Department at IFSTTAR (French Institute of Science and Technology for Transport, Development and Networks).
He is the author of many publications in the field of pathology, investigations, repair, strengthening, durability and management of bridges. He leads expertises on structures such as bridges, dams, nuclear power plants, and participates actively in the elaboration of the French and European technical doctrine. He is also Senior Lecturer at the Ecole Nationale des Ponts et Chaussées (ENPC) and is involved in several international associations like IABSE, RILEM, IABMAS.

Christophe PAGET

AIRBUS Operations Ltd, UK

PRESENTATION

Validation, Verification and Implementation of SHM at Airbus

Friday, July 11th, 8:30-9:00 - Auditorium 450

BACKROUND

Christophe Paget has received his Master of Engineering (Hons) in Mechatronics from the University of Valenciennes, France. He received his PhD in Electronics from the University of Valenciennes early 2001 and his PhD in Aeronautics from the Royal Institute of Technology of Stockholm late 2001. He worked for the Swedish Defence Research Agency (FOI), Stockholm, Sweden from 1996 till 2002 on structural health monitoring. He now works at the Materials and Processes department of Airbus in the UK, since March 2002, leading the A350 Materials & Processing Integration team, where he carried out his work on SHM to date. He received his honorary visiting Professorship from University of Cardiff, UK in 2011 for his contribution to SHM.

Clemens BOCKENHEIMER

AIRBUS Operations GmbH, Germany

PRESENTATION

Validation, Verification and Implementation of SHM at Airbus

Friday, July 11th, 8:30-9:00 - Auditorium 450

BACKROUND

Clemens Bockenheimer is working with Airbus since 2006 and is currently head of A350 Materials & Processes for Testing, Surface and Standardisation. In addition, he is leading the technology fields of Structural Health Monitoring and Non-Destructive Bond Quality Assessment / Extended Non-Destructive Testing. He has initiated and is maintaining various networks and cooperation to foster time- and cost-efficient technology development. Furthermore, he is executive board member of the Aerospace Industry Steering Committee for SHM (AISC-SHM) which has developed the first Standard document, the SAE ARP6461 (SHM Guidebook) in the field of Aerospace. He graduated as engineer in material science in 1999 at the University of Saarland in Saarbrücken/Germany and additionally qualified as European Adhesive Engineer at the Fraunhofer Institute for Applied Material Science Bremen/Germany in 2002. He carried out scientific research in the department of Prof. Dr. Wulff Possart, focusing on the characterisation of adhesion and interphase processes in aluminum-epoxy compounds and took a doctor degree in material science engineering in 2003. He has 15 years of professional experience, covering the fields of material science and engineering, non-destructive testing, research, technology and aircraft development as well as international cooperation.

Andreas JEROMIN

Project engineer - Forschungs- und Entwicklungszentrum FH Kiel GmbH, Germany

PRESENTATION

Structural Monitoring on Germanys Offshore Research Platform FINO3

Friday, July 11h, 9:00-9:30 - Auditorium 450

BACKROUND

Andreas Jeromin, born 1978, studied aeronautics and astronautics engineering at the University of Stuttgart and graduated in 2003.

Afterwards, he began as project engineer at the German Aerospace Center in Stuttgart. In the Institute of Combustion Technology he developed a numerical model to include film cooling in combustion chamber simulations.

Thermal Methods and Modeling
Organized by Jean Dumoulin (IFSTTAR, Nantes)
Chairman and co-chairman(s): Francesco Soldovieri (CNR-IREA, Italy) and Vincent Feuillet (Univ. Paris-Est, CERTES / OSU Efluve, France).
Tuesday, July 8th, 10:30-12:30 - Room I

Signal Processing for Monitoring of Wind Turbines
Organized by Laurent Mevel (Inria, France)
Chairman and co-chairman(s): Laurent Mevel (Inria, France) and Michael Döhler (Inria, France).
Tuesday, July 8th, 16:10-18:10 - Room 200

OFS-4-SHM (Optical Fiber Sensors for Structural Health Monitoring)
Organized by Pierre Ferdinand (CEA LIST, France) and George Akhras (Centre for Smart Materials and Structures, Royal Military College of Canada, Canada)
Chairman and co-chairman(s): Pierre Ferdinand (CEA LIST, France) and George Akhras (Centre for Smart Materials and Structures, Royal Military College of Canada, Canada).
Wednesday, July 9th, 10:30-12:30 - Auditorium 450

Probabilistic SHM 1
Organized by Daniele Zonta, (University of Trento, Italy)
Chairman and co-chairman(s): Daniele Zonta (University of Trento, Italy) and Franck Schoefs (Université de Nantes).
Wednesday, July 9th, 10:30-12:30 - Room GH

Probabilistic SHM 2
Organized by Branko Glisic (Princeton University, USA)
Chairman and co-chairman(s): Branko Glisic (Princeton University, USA) and Francesca Lanata (Ecole Supérieure du Bois, France).
Wednesday, July 9th, 14:00-16:00 - Room GH

Advances in Monitoring with Ultrasonic Coda Waves
Organized by Odile Abraham (IFSTTAR, France), Vincent Tournat (Laboratoire d'Acoustique de l'Université du Maine, France) and Ernst Niederleithinger (BAM Federal Institute for Materials Research and Testing, Germany)
Chairman and co-chairman(s): Vincent Tournat (Laboratoire d'Acoustique de l'Université du Maine, France) and Ernst Niederleithinger (BAM Federal Institute for Materials Research and Testing, Germany).
Wednesday, July 9th, 14:00-16:00 - Room 200

Achievements and Prospects for Nanotechnologies and Nanosensors in SHM
Organized by Bérengère Lebental (IFSTTAR, France)
Chairman and co-chairman(s): Bérengère Lebental (IFSTTAR, France) & Fulvio Michelis (IFSTTAR, France).
Thursday, July 10th, 10:30-12:10 - Room I

Structural Health Monitoring of Civil Engineering Structures
Organized by Yavuz Kaya (The University of British Columbia, Canada)
Chairman and co-chairman(s): Frédéric Bourquin (IFSTTAR, France) and Philippe Gueguen (ISTerre/UGA/CNRS/IFSTTAR, France).
Thursday, July 10th, 13:50-15:30 - Room KL

Structural Identification and Damage Detection using Vibration Measurements
Organized by Costas Papadimitriou (University of Thessaly, Greece)
Chairman and co-chairman(s): Costas Papadimitriou (University of Thessaly, Greece) and Chatzi Eleni (ETH-Zurich, Switzerland).
Friday, July 11th, 10:30-12:30 - Auditorium 450

Special Session: SHM & PHM in Action

Session Chair: C. Boller (Chair in NDT & QA, University of Saarland, Saarbrücken, Germany)

Thursday, July 10th in auditorium 450

Objective

SHM & PHM in action session is one of the highlighted events of EWSHM showcasing how Structural Health Monitoring works in practical applications. This session features demonstrations on a variety of SHM & PHM applications from many leading or start-up companies and educational institutions in order to

- show the audience how SHM and PHM work in practical applications
- a better understanding of the practical issues of different SHM and PHM systems
- a getting further feedback and requirements expressed from current and potential SHM and PHM users
- share experience in SHM and PHM

The targets of the session SHM & PHM in action

There is a significant progress in SHM and PHM technologies recently for a variety of applications. The session SHM & PHM in action focuses on how these SHM and PHM systems work in practice in terms of installation, handling, interpretation, and robustness for the following applications:

- operational loads monitoring
- damage detection
- health monitoring
- life cycle management

More precisely, the session SHM & PHM in action intends to show as much of demonstration cases as possible, addressing as many of the aspects mentioned below:

- the way the monitoring system and the test is operated
- type of sensors and actuators (if required) and their way of attachment to components, linkage to the signal generation and acquisition unit, etc.
- signal generation and acquisition unit as a hardware and how it operates
- the way input data are entered and sensor data are received and how the result is presented during the test
- procedure for sensor signal processing
- the 'man-machine interface' such as data input and output display
- component(s) tested, area/volume to be monitored, loading procedure and the damage initially observed by conventional means of non-destructive testing
- characteristics of the system such as weight, size, volume, reliability, cost, etc.

A detailed program of the session SHM & PHM in action is available in the conference kit.

Tuesday July 8, 2014

8:30-9:00
Auditorium TuOC

Vincent Le Cam (IFSTTAR), Laurent Mevel (Inria) and Franck Schoefs (Université de Nantes):
Opening

9:00-9:30
Auditorium TuKN1L

Fu Kuo CHANG (Stanford University):
The Need of SHM Quantification for Implementation

9:30-10:00
Auditorium TuKN2L

Christian BOLLER (University of Saarland):
Structural Health Monitoring - What WE Need to Do Now

Track T1	Track T2	Track T3	Track T4	Track T5	Track T6
10:30-12:30 **Auditorium** TuAT1 Aeronautics I	**10:30-12:30** **GH Room** TuAT2 Guided Waves I	**10:30-12:30** **Room I** TuAT3 Thermal Methods and Modeling	**10:30-12:30** **Room J** TuAT4 SHM for Railroads	**10:30-12:30** **KL Room** TuAT5 Sensors I	**10:30-12:30** **Room 200** TuAT6 Wind Energy I
13:50-15:50 **Auditorium** TuBT1 Aeronautics II	**13:50-15:50** **GH Room** TuBT2 Guided Waves II	**13:50-15:50** **Room I** TuBT3 Environmental Effects	**13:50-15:50** **Room J** TuBT4 Bridge Monitoring	**13:50-15:50** **KL Room** TuBT5 Sensors II	**13:50-15:50** **Room 200** TuBT6 Wind Energy II
16:10-18:10 **Auditorium** TuCT1 Composites I	**16:10-18:10** **GH Room** TuCT2 Guided Waves III	**16:10-18:10** **Room I** TuCT3 SHM Applications I	**16:10-18:10** **Room J** TuCT4 Image Processing	**16:10-18:10** **KL Room** TuCT5 Piezoelectric Sensing	**16:10-18:10** **Room 200** TuCT6 Signal Processing for Monitoring of Wind Turbines

18:10-20:00
Exhibition TuPoster

Poster Session + Cocktail Party in Exhibition Space

Wednesday July 9, 2014

8:30-9:00
Auditorium WeKN1L

Falk HOFFMANN (Airbus Helicopters):
Enabling Condition Based Maintenance for Helicopters

9:00-9:30
Auditorium WeKN2L

Pierre DERSIN (Alstom Transport):
An Attempt at Diagnosing the Present and Prognosticating the Future...

9:30-10:00
Auditorium WeKN3L

Claus-Peter Fritzen and Peter Kraemer (Siegen University and Woelfel Company):
Monitoring of Wind Energy Plants - Challenges and State of the Art

Track T1	Track T2	Track T3	Track T4	Track T5	Track T6
10:30-12:30 **Auditorium** WeAT1	**10:30-12:30** **GH Room** WeAT2	**10:30-12:30** **Room I** WeAT3	**10:30-12:30** **Room J** WeAT4	**10:30-12:30** **KL Room** WeAT5	**10:30-12:30** **Room 200** WeAT6
OFS-4-SHM (Optical Fiber Sensors for Structural Health Monitoring)	Probabilistic SHM 1	Sensors III	Pattern Recognition I	Structural Simulation	Acoustic Emission
14:00-16:00 **Auditorium** WeBT1	**14:00-16:00** **GH Room** WeBT2	**14:00-16:00** **Room I** WeBT3	**14:00-16:00** **Room J** WeBT4	**14:00-16:00** **KL Room** WeBT5	**14:00-16:00** **Room 200** WeBT6
Fiber Optics I	Probabilistic SHM 2	Sensors IV	Pattern Recognition II	FEM Methods and SHM	Monitoring with Ultrasonic Coda Waves
16:20-18:00 **Auditorium** WeCT1	**16:20-18:00** **GH Room** WeCT2	**16:20-18:00** **Room I** WeCT3	**16:20-18:00** **Room J** WeCT4	**16:20-18:00** **KL Room** WeCT5	**16:20-18:00** **Room 200** WeCT6
Fiber Optics II	Life Cycle Analysis	Realtime Methods	Pattern Recognition III	Modal Analysis	Nonlinear Acoustics

19:30-23:30

Gala dinner at les machines de l'île de Nantes in the Elephant Hall

Thursday July 10, 2014

8:30-9:00
Auditorium ThKN1L

Benjamin L'HENORET and Hughes GIGLEUX (SNCF Company):
Track Condition Monitoring at SNCF : An Introduction to Methods, Tools and Perpectives

9:00-9:30
Auditorium ThKN2L

Pierre STEPHAN and Jean SALIN (EDF Company):
Improvement of Civil Works Health Management : EDF's Global Strategy and Major Advances between 2008 and 2013

9:30-10:00
Auditorium ThKN3L

Carlos VENTURA (British Columbia University):
Structural Health Monitoring of Civil Engineering Structures - Lessons Learned and Future Directions

Track T1	Track T2	Track T3	Track T4	Track T5	Track T6
10:30-12:10 **Auditorium** ThAT1	**10:30-12:10** **GH Room** ThAT2	**10:30-12:10** **Room I** ThAT3	**10:30-12:10** **Room J** ThAT4	**10:30-12:10** **KL Room** ThAT5	**10:30-12:10** **Room 200** ThAT6
Strain Monitoring I	Statistical Approaches I	Achievements and Prospects for Nanotechnologies and Nanosensors in SHM	Electromagnetic SHM	Civil Engineering SHM	Damage I
13:50-15:30 **Auditorium** ThBT1	**13:50-15:30** **GH Room** ThBT2	**13:50-15:30** **Room I** ThBT3	**13:50-15:30** **Room J** ThBT4	**13:50-15:30** **KL Room** ThBT5	**13:50-15:30** **Room 200** ThBT6
Strain Monitoring II	Statistical Approaches II	SHM Applications II	Guided Waves IV	Monitoring of Civil Engineering Structures	Damage II

15:50-17:50
Auditorium

SHM + PHM in Action

18:45-23:30

Cruise ship on Erdre river and dinner at Le Chateau de la Poterie

Friday July 11, 2014

8:00-8:30 Auditorium FrKN1L	Frédéric BOURQUIN and Bruno GODART (IFSTTAR): **SHM Trends and Opportunities for the Civil Engineering Sector**
8:30-9:00 Auditorium FrKN2L	Christophe PAGET and Clemens BOCKENHEIMER (Airbus): **Validation, Verification and Implementation of SHM at Airbus**
9:00-9:30 Auditorium FrKN3L	Andreas JEROMIN (FINO3): **Structural Monitoring on Germanys Offshore Research Platform FINO3**
9:30-10:00 Auditorium FrKN4L	Vincent Le Cam (IFSTTAR), Laurent Mevel (Inria) and Franck Schoefs (Université de Nantes): **Closing**

Track T1	Track T2	Track T3	Track T4	Track T5	Track T6
10:30-12:30 Auditorium FrAT1	**10:30-12:30** GH Room FrAT2	**10:30-12:30** Room I FrAT3	**10:30-12:30** Room J FrAT4	**10:30-12:30** KL Room FrAT5	**10:30-12:30** Room 200 FrAT6
Structural Identification and Damage Detection Using Vibration Measurements	Composites II	SHM Applications III		Signal Processing	SHM Methodologies

14:00-17:00	External Visit by Cars **Industrial and R&D Sites Visit**

Tuesday July 8, 2014

Keynote Lectures	Auditorium
Chair: Le Cam, Vincent	IFSTTAR

09:00-09:30	TuKN1L

The Need of SHM Quantification for Implementation

Chang, Fu Kuo — Stanford University

Structural health monitoring (SHM) technology offers a new approach to interrogate the integrity of structures in real -time or on demand without physically disassembling the structures, unlike tradition al inspection techniques such as Non-destructive Inspection (NDI) whichrequire structural disassembly as well as a human to perform the inspection. SHM uses permanently mounted sensors to gather data from structures while typical NDI methods utilize removable probes to gather data. Both SHM and NDI may share the same physics for structural integrity detection during operation, however, there clearly exist fundamentally different engineering disciplines in these two approaches.

In order to transition SHM from R&D to field implementation, the reliability of the technique must be demonstrated and quantified using acceptable established procedures used by the industry. Probability of detection (POD) is a well accepted measurement method to quantify NDI results. This talk will examine the similarities and differences between traditional ultrasound methods and Lamb-wave based SHM techniques, and will identify key areas where reliability assessment of SHM may be different from traditional NDI. A procedure for evaluating the detection capability of SHM is proposed. It is anticipated that more studies will be needed to generate data to validate the procedure.

09:30-10:00	TuKN2L

Structural Health Monitoring – What WE Need to Do Now

Boller, Christian — Fraunhofer

Structural health monitoring (SHM) is now discussed and researched for various decades. The 'date of birth' of SHM is not very well specified. Some date it back to the Comet accidents in the early 1950ies, others date it back to when composite materials came into play in aviation which was more in the 1970ies or even later. SHM has been extremely much considered for aviation but the applications are still comparatively limited. SHM business cases for aviation look to be extremely limited in case design and maintenance philosophies will not be completely changed. Aviation structures as well as their maintenance today have gained such an incredible level of safety and reliability and have so much been optimized from a design point of view and risks taken have been put very low for good reasons, that SHM in aviation has a difficulty to show a business case under current conditions. Furthermore conventional NDT with the human being's flexibility in operation can provide a much bigger data sampling than SHM with its statically positioned transducers can do. If SHM therefore wants to be competitive it must be very well designed towards the tolerable damage it has to detect reliably. This can only be done by thorough simulation before realizing an optimized SHM system. Today possibly myriads of algorithms and simulations are performed in SHM wordwide. They are all circumventing similar physical damage mechanism phenomena. But is this possibly most essential instrument for SHM truly linked together? Are WE truly communicating here as we do with all other means of communication today? The presentation will address this aspect being also linked to a project recently launched, which has only been considered as a nucleus to a much bigger task.

TuAT1 **Aeronautics I** (Regular Session)	Auditorium
Chair: Holnicki-Szulc, Jan	IFTR Warsaw
Co-Chair: Gogu, Christian	Univ. of Toulouse

10:30-10:50	TuAT1.1

Monitoring of a Full-Scale Wing Fatigue Test

Heida, Jaap	National Aerospace Lab. NLR
Hwang, Jason	National Aerospace Lab. NLR

A wing of a decommissioned aircraft of the Royal Netherlands Air Force (RNLAF) was fatigue tested to more than two times the design life. Part of the test was the evaluation of load monitoring and Structural Health Monitoring (SHM) techniques. For load monitoring the data of conventional resistance strain gauges was compared with the response of optical Fibre Bragg Gratings (FBG). For SHM the Acoustic Emission (AE) and Comparative Vacuum Monitoring (CVM) techniques were employed to monitor fatigue crack initiation and growth. With the optical fibres a linear correlation between the FBG's and the conventional strain gauges was obtained using a specific strain transfer factor. The SHM techniques, on the other hand, were less successful. For the CVM technique no cracks occurred at the locations under periodic monitoring (but also no false calls occurred during the complete test). Further, the AE system under continuous monitoring registered a lot of AE activity from different sources (also after drastic filtering of the AE data) but the AE data could not be reliably related to the initiation and growth of fatigue cracks in the areas monitored. Most of the AE activity was probably caused by mechanically induced noise such as frictional noise from the fastener locations and other surface rubbing areas.

10:50-11:10	TuAT1.2

Integration of Embedded Data Processing Algorithms Inside PAMELA Devices

Ruiz, Mariano	Univ. Pol. de Madrid
Barrera, Eduardo	Tech. Univ. of Madrid
Fernandez, Nestor	Tech. Univ. of Madrid
Melendez, Raul	Tech. Univ. of Madrid
Alcaide Pardo, Ángel	AERNNOVA Engineering Solutions Iberica S.A.
Aranguren, Gerardo	UPV/EHU
Monje, Pedro M.	Electronic Design Group, Univ. of the Basque Country

PAMELA (Phased Array Monitoring for Enhanced Life Assessment) SHM System is an integrated embedded ultrasonic guided waves based system consisting of several electronic devices and one system controller. The data collected by all PAMELA devices in the system must be transmitted to the controller, who will be responsible for carrying out the advanced signal processing to obtain SHM maps. PAMELA devices consist of hardware based on a Virtex 5 FPGA with a PowerPC 440 running an embedded Linux distribution. Therefore, PAMELA devices, in addition to the capability of performing tests and transmitting the collected data to the controller, have the capability of perform local data processing or pre-processing (reduction, normalization, pattern recognition, feature extraction, etc.). Local data processing decreases the data traffic over the mesh network and allows CPU load of the external computer to be reduced. Even it is possible that PAMELA devices are running autonomously performing scheduled tests, and only communicates with the controller in case of detection of structural damages or when programmed. Each PAMELA device integrates a software management application that allows to the developer to download his own algorithm code and to add the new data processing algorithm to the device. The development application is done in a virtual machine with an Ubuntu distribution including all necessary software tools to perform the entire cycle of development. Eclipse IDE (Integrated Development Environment) is used to develop the project and to write the code of each data processing algorithm. This paper presents the developed software architecture and describes the necessary steps to integrate new embedded data processing algorithms inside PAMELA devices.

11:10-11:30 TuAT1.3

Full Field Stress Measurement for in Situ Structural Health Monitoring of Airframe Components and Repairs

Rajic, Nik — Defence Science and Tech. Organisation
Galea, Steve — Defence Science and Tech. Organisation

The fatigue usage monitoring systems installed on various military aircraft rely primarily on strain gauges for sensory information, and for good reason. Strain gauges have a well established certification framework, a relatively good track record of reliability and they directly target the parameter that drives fatigue. Extending the role of strain gauges to structural health monitoring however is problematic. The reasons are manifold but a key one is that strain gradients at ``fatigue hot spots" are often of a length scale shorter than the gauge. Stress imaging techniques have no such restriction but few have realistic prospects for application to in situ structural health monitoring. Recent developments in miniature low-cost microbolometer technology could make thermoelastic stress analysis an exception. The present paper examines the potential of this class of infrared detector to furnish a basis for in situ structural health monitoring. It first covers some preliminaries starting with a brief comparison of thermoelastic stress analysis to other full-field strain measurement techniques, followed by a discussion of several aircraft related applications illustrating the key properties of the methodology. Finally, it describes a case study in structural health monitoring involving the centre barrel assembly of an F/A-18 A/B fighter aircraft.

11:30-11:50 TuAT1.4

Rapidly Building a Parametric Model to Estimate the Structural Fatigue of a Fast Jet

Waldock, Antony — BAE Systems
Hunt, Stephen — BAE Systems
Kelsey, Glen — BAE Systems

To effectively plan maintenance actions for a fleet of fast jets requires accurately monitoring the fatigue of the structure. In this application, a parametric model is used to estimate the stress experienced at different locations on the aircraft given flight parameters (acceleration forces, speed etc.) collected from on-board sub-systems. The estimated stress is then combined with previous stress cycles to calculate the overall fatigue of the structure at different locations. The problem addressed in this paper, is the rapid generation of new parametric models that accurately estimate the stress experienced when new aircraft configurations are introduced. Firstly, an initial parametric model is constructed using a physics-based model of the new aircraft configuration. Secondly, and the main focus of this paper, a tool based on numerical optimisation methods is used to refine the parametric model using data collected from test flights, where flight parameters and strain measures at different locations are recorded on a test aircraft in the new configuration. This paper presents how an optimisation tool can be utilised by a structural engineer to refine an existing parametric model to estimate the structural fatigue of the new aircraft configuration and presents a brief analysis of different parameters within the tool. The paper concludes that the tool enables a parametric model to be rapidly generated that accurately estimates fatigue, and hence enables the maintenance procedures to be scheduled for new aircraft configurations.

11:50-12:10 TuAT1.5

Challenges of an Industrialized Acousto-Ultrasonic Sensor System Installation on Aircraft Primary Structure

Dobmann, Nicolas — EADS Innovation Works
Bach, Martin — EADS Deutschland GmbH
Eckstein, Benjamin — EADS Innovation Works

At laboratory level, secondary bonding of Acousto-Ultrasonic sensor systems is well established and has proven its reliability in applications from coupon level up to flight test installations. However, the applied sensor secondary bonding is a manual process with high amount of required auxiliaries and tools and is hence associated with high costs. In transition from sensor installation under laboratory conditions to the installation of large-scale sensor networks within an aircraft serial production process, costs associated with sensor installation become a decisive factor. By developing the sensor co-bonding approach, an installation method for fiber reinforced plastics host structures is created which requires significantly less work, auxiliaries and tools. Yet, while a considerable reduction of installation cost seems achievable, the application of sensor co-bonding as an industrialized process for aircraft primary structure is clearly more challenging. The present paper outlines specific advantages of sensor co-bonding and discusses challenges ahead on the way towards an industrialized sensor installation on aircraft primary structures.

12:10-12:30 TuAT1.6

Detectability Assessment of Optical Fiber Sensor Based Impact Damage Detection for Composite Airframe Structures

Hirano, Noriyoshi — Kawasaki Heavy Industries, LTD

Airframe structures, especially composite structures, are prone to damage by tool drops, bird strikes, hailstones, etc. The authors have developed an optical fiber based impact damage detection system for composite airframe structure since 1998. In this method, impact locations are measured with the signals of multiple Fiber Bragg Grating (FBG) sensors. In case of aircraft implementation, impact information should be detected under the actual vibration condition, and both detection capability for impact localization and damage level should be improved. This paper presents the detectability assessment result from the impact tests under a simulated vibration condition. In addition, study of optimal FBG sensor direction for localization and review of damage level detection method are conducted to enhance the accuracy of detection.

TuAT2 — GH Room

Guided Waves I (Regular Session)

Chair: Stargel, David — US Airforce
Co-Chair: Lecce, Leonardo — Univ. of Naples Federico II

10:30-10:50 TuAT2.1

An On-Line Continuous Updating Gaussian Mixture Model for Damage Monitoring under Time-Varying Structural Boundary Condition

Qiu, Lei — Nanjing Univ. of Aeronautics and Astronautics
Yuan, Shenfang — Nanjing Univ. of Aeronautics and Astronautics
Bao, Qiao — Nanjing Univ. of Aeronautics and Astronautics
Huang, Tianxiang — Nanjing Univ. of Aeronautics and Astronautics

Damage monitoring under time-varying structural boundary condition is one of the most difficult tasks in piezoelectric transducers (PZTs) and Lamb wave based SHM methods for engineering applications. Because the structural boundary changes such as variations in the tightness of bolts between structures can lead to false monitoring result even the structure is in health state. This paper proposes a Lamb wave based on-line continuous updating Gaussian Mixture Model (GMM) to study the problem. Based on the baseline GMM constructed by features of Lamb wave signals in structural health state, an on-line continuous updating GMM is studied to learn the dynamic changes of Lamb wave monitoring signals without any prior knowledge of damage patterns. The Kullback–Leibler (KL) divergence is used as a degradation index to estimate the structural damage by measuring the difference between the baseline GMM and the on-line GMM. The proposed method is validated on an aircraft steel beam. The validation results show that the method is effective for bolt hole crack growth monitoring under the time-varying changes in the tightness degree of the bolts.

10:50-11:10 TuAT2.2

Analytically-Based Simulation for Corrosion Detection by Guided

Waves

Glushkov, Evgeny	Kuban State Univ. Inst. for Mathematics, Mechanics and
Glushkova, Natalia	Kuban State Univ. Inst. for Mathematics, Mechanics and
Eremin, Artem	Kuban State Univ.
Giurgiutiu, Victor	Univ. of South Carolina

To create a theoretical basis for guided wave detection and identification of corrosion damages, a set of analytically based computer models of various complexity has been developed. The present paper is focused on the simplest and fastest beam model for stepped and notched waveguides, which has exhibited a wide frequency range of reasonable coincidence with the results obtained within more complex integral equation based model for a 2D notched elastic strip.

11:10-11:30 TuAT2.3

The Spectral Cell Method for Ultrasonic Guided Wave Propagation Problems

Duczek, Sascha	Otto-von-Guericke-Univ. Magdeburg
Liefold, Steffen	Otto-von-Guericke-Univ. Magdeburg
Gabbert, Ulrich	Otto-von-Guericke-Univ. Magdeburg, Inst. für Mechanik

In the current paper we present a fast and robust numerical tool for the simulation of ultrasonic guided waves in heterogeneous structures. The proposed approach, the so-called *spectral cell method* (SCM), combines the fundamental ideas of the spectral element method (SEM) with the fictitious domain concept. The SCM accordingly retains the high convergence rates known from high-order finite element methods and circumvents the need for body-fitted discretizations. Mass-lumping techniques being available for the SEM can also be applied for the SCM, which offers benefits when explicit time integration methods such as the central difference method (CDM) are employed. Due to these properties both memory requirements and computational time can be notably reduced. The SCM therefore paves the way for an efficient simulation of ultrasonic guided waves. In the first part of the paper we introduce the basic principles of high-order finite element methods (for multi-physics applications - piezoelectricity) and the fictitious domain approach to illustrate the behaviour of the proposed method. The second part contains numerical examples showing that the performance of the SCM is comparable to other (established) high-order methods.

11:30-11:50 TuAT2.4

Characterization of Laminar Damage in an Aluminium Panel by Diffraction Tomography Based Imaging Method Using Lamb Waves

Chan, Eugene	RMIT Univ.
Wang, Chun	RMIT Univ.
Rose, Francis	RMIT Univ.

Most of the research on Structural Health Monitoring employs guided waves only to detect and locate damage in a plate-like structure. The purpose of this paper is to present a Diffraction Tomography based imaging method, using experimentally-determined scattered fields (Lamb wave signals) and numerically computed Green's function, to characterise laminar damage in an aluminium panel. The approach is based on a recently derived extension of Diffraction Tomography which utilises the multi-static scattering matrix constructed from the measurements of the scattered field for every source and receiver pair, as well as the Green's function of the structure which is its response to a point source. The imaging results have provided for the first time an accurate characterisation of damage geometry and size derived from experimental data. These results are shown to compare favourably with those obtained from computational data, and they are significantly more accurate than previously reported results.

11:50-12:10 TuAT2.5

Guided Waves in Layered Plate with Delaminations

Ricci, Fabrizio	Univ. of Naples Federico II
Mal, Ajit	Univ. of California, Los Angeles
Monaco, Ernesto	Department of Industrial Engineering - Univ. of Naples Fede
Maio, Leandro	Department of Industrial Engineering - Univ. of Naples Fede
Boffa, Natalino Daniele	Department of Industrial Engineering - Univ. of Naples Fede
Di Palma, Marco	Department of Industrial Engineering - Univ. of Naples Fede
Lecce, Leonardo	Univ. of Naples Federico II

The integrity of safety-critical structural composites can be enhanced by the use of innovative ultrasonic nondestructive evaluation (NDE) techniques. Among the various existing techniques, guided wave methods provide a good compromise in terms of sensitivity to a variety of damage types or defects and extent of the area that can be monitored, given the ability of this waves to travel relatively long distances within the structure under investigation. In comparison with Lamb waves propagating in isotropic structures, wave propagation in composite structures presents additional complexity for effective damage identification. The material inhomogeneity, the anisotropy and the multi-layered construction lead to the significant dependence of wave modes on laminate layup configurations, direction of propagation, frequency, and interface conditions. In this paper a specific structure will be analyzed with different levels of complexities as far as the wave propagation characteristics are concerned. The investigated structure is a composite stiffened plate. The work is carried out using theoretical analysis, numerical models and experimental verifications. Numerical (Finite Element) models are used for more practical cases, for which the geometric and material complexities of actual structures present practical difficulties in direct analysis of wave propagation data using theoretical constructs only.

TuAT3 Room I

Thermal Methods and Modeling (Invited Session)

Chair: Soldovieri, Francesco	CNR-IREA
Co-Chair: Feuillet, Vincent	Univ. Paris-Est, CERTES / OSU Efluve

10:30-10:50 TuAT3.1

Active Thermal Shearography and Infrared Thermography Applied to NDT of Reinforced Concrete Structure by Glued CFRP (I)

Theroux, Louis-Daniel	IFSTTAR
Dumoulin, Jean	IFSTTAR
Maldague, Xavier	Univ. Laval

This research paper presents the study of thermography and shearography to evaluate CFRP reinforcement of concrete structures. The study explores the strengths and weaknesses of both non-destructive-testing (NDT) methods. It will be shown that by coupling the methods, the detection is more reliable and the defect evaluation more thorough as both the thermal and thermo mechanical properties are tested. Once the core theoretical concept regarding both methods is presented, the viability is demonstrated in the experimentation part of this study. With the experimental results confirming the feasibility of the coupling of both methods, the numerical model conceived allow to better conduct analysis that are not otherwise possible with the experimental results.

10:50-11:10 TuAT3.2

Diagnosis of Insulated Building Walls Using Passive Infrared Thermography and Numerical Simulations (I)

Monchau, Jean-Pierre	Univ. Paris-Est, CERTES
Ibos, Laurent	Univ. Paris12
Feuillet, Vincent	Univ. Paris-Est

This work presents the thermal monitoring of a multi-layered wall of a restored building (PANISSE platform). Surface temperatures measured by infrared thermography are corrected by taking account of the influence parameters. Then they are compared with thermocouple measurements and numerical simulations.

11:10-11:30 TuAT3.3

Remocean: A X-Band Radar System As Tool for Environmental Monitoring in Costa Concordia Ship Wreck Management (I)

Arturi, Daniele — Univ. MEDITERRANEA
Soldovieri, Francesco — CNR-IREA
Natale, Antonio — IREA-CNR
Lugni, Claudio — INSEAN-CNR
Ludeno, Giovanni — IREA--CNR
Serafino, Francesco — CNR

This paper describes the new opportunities made possible by the X-band radar systems for monitoring and management of infrastructures in coastal areas as well as for offshore platforms. In particular, the paper presents the effectiveness and the reliability of the Remocean system, a X-band radar developed at IREA-CNR, as observational tool for the support of the Costa Concordia ship wreck removal and as a scientific opportunity to observe wave phenomena rarely observable in deep sea.

11:30-11:50 TuAT3.4

Association of Sensing Techniques with a Designed ICT Architecture in the ISTIMES Project: Application Example with the Monitoring of the Musmeci Bridge (I)

Soldovieri, Francesco — CNR-IREA
Dumoulin, Jean — IFSTTAR
Ponzo, Felice Carlo — Univ. of Basilicata
Crinière, Antoine — IFSTTAR
Bourquin, Frédéric — IFSTTAR
Cuomo, Vincenzo — CNR-IMAA

This work gives a brief description of the main activities and outcomes of the Integrated System for Transport Infrastructures surveillance and Monitoring by Electromagnetic Sensing (ISTIMES) project, which aimed at designing and implementing a system able to couple the capabilities of long-term monitoring and quick damage assessment of the critical transport infrastructures. This was performed thanks to the integrated use of the novel and state of art concepts of Earth observation, ground-based sensing techniques and ICT architecture. The paper will give a brief outline of the main results of the project by referring in particular to the demonstration activities at the test bed of the "Musmeci" Bridge in Potenza, Southern Italy.

11:50-12:10 TuAT3.5

Radar Systems for Infrastructures Diagnostics: A Review (I)

Catapano, Ilaria — Inst. for Electromagnetic Sensing of the Environment (IREA)
Affinito, Antonio — CNR-IREA
Crocco, Lorenzo — CNR-IREA
Soldovieri, Francesco — CNR-IREA

This communication aims at summarizing the main features, recent advancements and examples assessing the reconstruction capabilities of two different radar systems for subsurface imaging and civil engineering monitoring and diagnostics. In fact, Ground Penetrating Radar and Holographic Radar are well assessed non-invasive tools capable of investigating the inner status of an object and are attracting more and more attention in several applicative scenarios, for their easiness of use and significant imaging capabilities.

12:10-12:30 TuAT3.6

ReTrofiT: A Generic Software to Solve Optimization and Identification Problems Applied to Building Energy Management (I)

Nassiopoulos, Alex — LUNAM Univ. IFSTTAR, 44344, Bouguenais, France
Brouns, Jordan — LUNAM Univ. IFSTTAR, 44344, Bouguenais, France
Azerou, Boussad — LUNAM Univ. IFSTTAR, 44344, Bouguenais, France
Artiges, Nils — LUNAM Univ. IFSTTAR, 44344, Bouguenais, France
Smail, Mostafa Kamel — IFSTTAR

Energy management systems in buildings greatly contribute to the improvement of overall energy efficiency. Monitoring systems can lead to significant reductions of the global energy use by increasing occupants' awareness of the consumptions or by enabling the implementation of more efficient regulation strategies. Model predictive control consists in computing optimal heating or cooling strategies by taking into account the future evolution of the state of the building under forecast weather or use conditions. Demand response strategies in smart grids consist in adjusting energy demand at the end-user level to reduce the overall demand thus resulting in end-user customer bill savings, increase of electricity market stability and of electricity supply reliability. Further, today the building construction practices evolve towards a more performance-based approach in which the concern becomes the performance of the final building rather than the means employed to construct it.

All the aforementionned applications rely on the ability to accurately predict a building's behavior using a calibrated model. In building energy applications, uncertainties in input data of modelling tools often lead to important discrepancies between the model predictions and the real performance. The desired model response can be obtained if the internal parameters of the model are calibrated using on-site measurements and model identification methods.

The paper presents the software ReTrofiT that was specifically designed to treat this kind of problems. ReTrofiT is first of all a dynamic building simulation code with multizone-type assumptions. It integrates a set of tools and algorithms to set up and solve minimization problems as well as to compute sensitivities. All these operations can be done with a negligible computation by means of the adjoint model that is intrinsically implemented.

TuAT4 — Room J

SHM for Railroads (Regular Session)

Chair: Bouillaut, Laurent — IFSTTAR
Co-Chair: Gigleux, Hugues — SNCF

10:30-10:50 TuAT4.1

Three Experimental Applications of Health Algorithms to Improve Infrastructure Inspection

Ervin, Elizabeth K. — The Univ. of Mississippi

This paper presents three applications of structural health evaluation from data collection to damage location. The first test setup was a standard railway track with an adjustable sleeper to allow for artificial defects. The second test setup was a student-built reinforced concrete bridge with various rubber bearings to represent joint softening. The third test location was a skewed on-campus highway bridge built in 1940 with no maintenance record. The obtained data sets were analyzed with an in-house structural health evaluation program, performing modal decomposition and applying twelve different damage detection algorithms.

10:50-11:10 TuAT4.2

Online Damage Monitoring for High-Speed Train Bogie Using Guided Waves: Development and Validation

Wang, Qiang — Nanjing Univ. of Posts and Telecommunications
Su, Zhongqing — The Hong Kong Pol. Univ.
Hong, Ming — The Hong Kong Pol. Univ.

The safety of high-speed trains has therefore become a key concern not only in the design process but also in their operation. Usually, plenty of testing jobs are performed through periodical maintenance to ensure the integrity of train structures, using offline nondestructive evaluation methods with relatively low cost-effectiveness. Based on the group's efforts in the past ten years, a guided wave based damage detection and monitoring technique was developed to provide an online structural damage inspection approach, aiming to increase the safety of bogie structures and improve train operation efficiency. Miniaturized standard PZT sensors were developed to compose a pitch-catch based active sensor network for guided wave excitation and acquisition in the train bogie, and a compact system built with the proposed technique was implemented online to inquire information on structural health conditions. As a part of the conformance testing of China's latest high-speed train model, experiments on a bogie frame of the train were carried out, especially when the train was running at a high speed, to validate the proposed technique and system, taking into account the complicated working states and the highly variable circumstances of the train, which are usually hard to be simulated in the lab. Several practical experiments were involved in the testing, including the survival rate of the sensor network, disturbance of mechanical vibration to signals, influences of actions of train (such as urgency brake), artificial damage detection, and so on. After running and testing for more than 1500 km, the experimental results from different conditions demonstrated high reliability and accuracy of the technique and the system.

11:10-11:30 TuAT4.3

A Fibre Optic Sensor Instrumented Pantograph As Part of a Continuous Structural Health Monitoring System for Railway Overhead Lines

Wagner, Richard	HBM
Maicz, Dietmar	Hottinger Baldwin Messtechnik GmbH
Saliger, Christian	MELECS MWW GmbH & Co KG
Noack, Thomas	TÜV SÜD Rail GmbH
Viel, Wolfgang	Hottinger Baldwin Messtechnik GmbH (HBM)
Saliger, Florian	ÖBB Infrastructure AG - Res. and Developent
Horak, Roman	MELECS MWW GmbH & CoKG

This article presents the project SAFE onLine (Pantograph for fibre-optic based recording and online evaluation of the catenary condition) funded by the Austrian Research Promotion Agency (FFG). The goal of this project is to develop a measurement system based on a commercial pantograph, that a regularly railway engine can be equipped with. We will give an overview on the different aspects and further course of the project and present first measuring results with the system obtained during integration tests at the pantograph test-bed.

11:30-11:50 TuAT4.4

The Selection Procedure of Diagnostic Indicator of Suspension Fault Modes for the Rail Vehicles Monitoring System

Melnik, Rafał	Warsaw Univ. of Tech.
Sowiński, Bogdan	Warsaw Univ. of Tech. Faculty of Transport

The monitoring system of a rail vehicle and track has been developed under the project MONIT – Monitoring of Technical State of Construction and Evaluation of its Lifespan. The main subsystem of a rail vehicle being the object of on-line monitoring is the suspension. The principle of operation of the monitoring system is based on the statistical analysis of acceleration signals. The selection procedure of damage-sensitive parameters of signals is introduced. In order to improve suspension condition assessment, selected parameters are analysed in multidimensional space. As a result, a new diagnostic indicator is analysed – Euclidean metric between reference point (nominal state) and a point representing current measurement (condition). The proposed fault detection method was used for analysis of recorded data during an experiment on wagons with suspension damages.

11:50-12:10 TuAT4.5

The Use of Acceleration Signals Recorded on a Railway Vehicle Wheelsets for Rail Track Condition Monitoring

Kostrzewski, Mariusz	Warsaw Univ. of Tech.
Chudzikiewicz, Andrzej	Warsaw Univ. of Tech. Faculty of Transport
Bogacz, Roman	Pol. Acad. of Science, Department of Intelligent Tech.

The subject matter of the paper is the partial analysis of the results of acceleration signals from the prototype of *Rail Vehicle's and Rail Track Monitoring System* exploitation The prototype of the system measures and records acceleration signals on some elements of electric multiple unit. The chosen analysis are connected to track condition, however connection to vehicle condition are also mentioned. Appropriate conclusions and future research proposal are given.

12:10-12:30 TuAT4.6

Monitoring of the Structural Integrity of Wheelset Axles Using Guided Waves

Grzeszkowski, Mateusz	Federal Inst. for Materials Res. and Testing
Prager, Jens	Federal Inst. for Materials Res. and Testing

The structural integrity of wheelset axles of high speed trains as well as freight wagons is of major importance for safety of rail transport. Up to now the inspection of wheelset axles is part of a periodic maintenance which is expensive due to long inspection times and highly influenced by the human factor, especially if those inspections are carried out during night time. Permanently installed SHM-systems at each axle would be able to monitor the structural integrity and could reduce the maintenance effort. However due to mechanical constraints, requirements and regulations the mounting options of such SHM-systems are limited. Approaches with ultrasonic transducers mounted at one end face of the axle can excite guided waves and could be able to monitor the entire component from only one sensor position. The geometry of the axle with many cross-sectional variations results in reflections and changing dispersion relations along the wave propagation direction and lead to complex wave propagation problems hiding flaw echoes. Thus, echo signals are difficult to interpret and correlations between echo signals and flaws are hard to reveal. In the presented paper we show a signal processing approach to extract information on the structural integrity from ultrasonic guided wave echo data. On a mock-up we could show, that the suggested method is able to detect a crack growth reliably. Based on the knowledge of wave propagation and on results from numerical simulations, classification algorithms are developed and applied to additionally enhance the prediction accuracy of crack size and position.

TuAT5 KL Room

Sensors I (Regular Session)

Chair: Billeres, Malvina	CEA
Co-Chair: Michelis, Fulvio	IFSTTAR

10:30-10:50 TuAT5.1

Chipless Passive Sensor for Wireless Monitoring of High Radiation Doses in Nuclear Infrastructures

Pons, Patrick	LAAS-CNRS
Aubert, Hervé	LAAS-CNRS
Debourg, Emilie	LAAS-CNRS
Rifai, Ayoub	LAAS-CNRS
Olszacki, Michal	National Centre for Nuclear Res.
Matusiak, Michal	National Centre for Nuclear Res.
Augustyniak, Isabella	Wrocław Univ. of Tech.
Knapkiewicz, Pawel	Wrocław Univ. of Tech.
Dziuban, Jan	Wrocław Univ. of Tech.
Lavielle, Denis	TRAD
Chatry, Christian	TRAD

The dosimetry is one of the crucial techniques that are needed to assure personal safety and facilities security in the areas of high radioactivity as nuclear power plants or powerful experimental research infrastructure (LHC, XFEL, ITER). Until now, the common electronic dosimeters are based on the silicon diodes or field effect transistors what limits the measured dose value up to few Mrad, as the higher levels lead to the device saturation. The other techniques that allows for higher dose measurement (up to 100 Mrad and more) are based mostly on the variation of physic-chemical parameters of different materials but need the post measurement treatment of the sensor that has to be taken out of the monitored zone and use of the special equipment what makes the continuous measurement impossible. Moreover, in case of nuclear reactors the access to the radioactive zones is limited what leads to the measurement frequency in range of a few times per year. The solution that may overcome that problem is the use of the recently developed innovative concept of the passive sensors that are interrogated by the radar technique. Such a solution uses the small (millimeter sized) micro-resonators designed for the ultra high frequency (about 30GHz) that are connected to the antenna and serve as its load. Any change of the resonant frequency of the resonator may be detected as the shift in the radar echo of the antenna. The sensor is interrogated by the directional radar and the reflected signal is analyzed. The associate transduction principle is that the resonator changes its resonant frequency due to the variation of the specific measured physical value like temperature, pressure or humidity. In our case, we proposed to use the known principle of Hydrogen-pressure dosimeters (HPD) wh

10:50-11:10	TuAT5.2

Structural Health Monitoring of Reinforced Concrete Beam Using Piezoelectric Energy Harvesting System

Cahill, Paul	Dynamical Systems and Risk Lab. Department of Civil and E
O'Keeffe, Rosemary	Microsystems Group, Tyndall National Inst. Univ. Coll.
Jackson, Nathan	Microsystems Group, Tyndall National Inst. Univ. Coll.
Mathewson, Alan	Microsystems Group, Tyndall National Inst. Univ. Coll.
Pakrashi, Vikram	Univ. Coll. Cork

There has been focus in recent times in the creation of smart, wireless sensor networks for the purposes of Structural Health Monitoring of large scale civil infrastructure. However, the power requirements of such networks are dependent on finite batteries, which limit the effectiveness of such a system. The use of energy harvesters, however, offers a viable and attractive solution to this problem. This paper investigates the use of such energy harvesters not only to power wireless sensor nodes, but to also act in the process as a damage detection tool. The properties and creation of such energy harvesters is detailed in full. The effects of damage on a simply supported reinforced concrete beam are investigated through finite element analysis. The use of the energy harvesters for damage detection is subsequently investigated and the feasibility of using such harvesters is experimentally validated. The simultaneous power of wireless sensor nodes by the harvesters is determined and an energy harvesting circuit is examined in this regard. This paper establishes the basis and viability of using an energy harvesting system for use in this dual role.

11:10-11:30	TuAT5.3

Passive Sensor Network Localization for Structural Health Monitoring

Rémy, Vincent	CEA, Leti, Grenoble, France.
Mikael, Carmona	CEA, Leti, Grenoble, France.
Olivier, Michel	Gipsa-Lab. Grenoble, France
Jean-Louis, Lacoume	Gipsa-Lab. Grenoble, France

In this article, passive acoustic sensor network localization is presented and applied to geometry structure monitoring. The methodology relies on passive travel-time estimation, which is here defined as the retrieval of an inter-sensor propagation delay, using uncontrolled ambient sources in an homogeneous acoustic propagation medium. Our approach relates to passive linear systems identification through the use of codas correlations to form estimators. We provide practical performances of such estimators and propose two approaches for geometry monitoring, both illustrated in the case of a steel beam.

11:30-11:50	TuAT5.4

Development of a Piezoelectric Based Energy Harvesting System for Autonomous Wireless Sensor Nodes

Gomes-Casseres Espinosa, Andrés Felipe	Univ. Distrital Francisco Jose de Caldas
Sanchez Ramirez, Andrea	Univ. of Twente
Loendersloot, Richard	Univ. of Twente
Berkhoff, Arthur	Univ. of Twente
Combita Alfonso, Luis Francisco	Univ. Distrital Francisco Jose de Caldas

This paper describes the selection and operation of a Boost Integrated with Flyback Rectifier/Energy storage/DC-DC converter (BIFRED) for piezoelectric energy harvesting purposes. This topology presents features like low-harmonic rectification, energy storage and wide-bandwith voltage control in an integrated single converter, leading to fewer components, weight and miniaturization of the entire system. This miniaturization makes the integrated topologies a suitable option for further developments in Wireless Sensor Network based Structural Health and Condition Monitoring applications. The operation of the BIFRED circuit in Discontinuous Conduction Mode (DCM) mode is described. A nonlinear model is derived and used to predict the DC gains between the input, storage and output voltages of the circuit. A simulation-based verification is carried out using MATLAB Simulink, the operation of the circuit is corroborated and the power factor is computed. Finally, the results of a piezoelectric implementation using the BIFRED circuit are presented.

11:50-12:10	TuAT5.5

Passive Magnetic Observer in NDE & SHM Applications

Witos, Miroslaw	Air Force Inst. of Tech.
Zokowski, Mariusz	Air Force Inst. of Tech.

The paper concerns about the idea of passive magnetic observer and magneto-mechanical effects (reversible and irreversible) to non-destructive testing and monitoring of critical elements (NDE&SHM). The diagnostic task is regarding as the fast identification of overload elements, which have not broken, cracked or deformed yet (detect 1st and 2nd stage of damage). Theoretical and experimental results from evaluating the early phase of the material fatigue have been presented. Issues have been discussed on two examples (classes of diagnosis task for passive magnetic observer): a) testing object is located only in the Earth's magnetic field; b) testing object is located in the Earth's magnetic field and low alternating electromagnetic field.

TuAT6	Room 200
Wind Energy I (Regular Session)	
Chair: Fassois, Spilios D.	Univ. of Patras
Co-Chair: Petryna, Yuri	Tech. Univ. Berlin

10:30-10:50	TuAT6.1

Lptv Subspace Analysis of Wind Turbines Data

Mevel, Laurent	Inria
Guéguen, Ivan	IFSTTAR
Tcherniak, Dmitri	Bruel and Kjaer Sound and Vibration Measurement

The modal analysis of a wind turbine has been generally handled with the assumption that this structure can be accurately modeled as linear time-invariant. Such assumption may be misleading for stability analysis, especially, with the current development of very large wind turbines with complex dynamic behavior (nonlinearity, aeroelastic coupling). Therefore in this paper, the inherent periodically time-varying dynamics of wind turbines (and for rotating systems, in general) is taken into account. Recently a subspace algorithm for modal analysis of rotating systems has been proposed. It is tested on a simulated and real data from a wind turbine.

10:50-11:10 TuAT6.2

Subspace-Based Detection of Fatigue Damage on Jacket Support Structures of Offshore Wind Turbines

Hille, Falk — BAM Federal Inst. for Materials Res. and Testing
Petryna, Yuri — Tech. Univ. Berlin
Rücker, Werner — BAM Federal Inst. for Materials Res. and Testing

The contribution describes the application of the Stochastic Subspace-based Damage Detection (SSDD) method on model structures for an utilization of this approach on offshore wind turbine structures. Aim of the study was therefore to analyze the usability and efficiency of the detection method as well as to determine an optimized set of parameter for realistic damage on support structures of wind energy turbines. Based on results of an experimental fatigue test on a steel frame laboratory structure a strategy for a numerical verification of the experimentally evolved damage detection was developed, utilizing a time integration approach to simulate the dynamic response. In a second step the identified modeling and computing methodology is used to numerically investigate the ability to detect damage in real size structural components of offshore wind turbines.

11:10-11:30 TuAT6.3

Damage Detection in a Wind Turbine Blade Based on Time Series Methods

Hoell, Simon — LRF Centre for Safety and Reliability Engineering, Univ. of
Omenzetter, Piotr — Univ. of Aberdeen

The interest in renewable energy in the European Union has increased in the past years, thus efficient energy harvesting becomes more important. For the sector of wind energy, the consequences are growing sizes of wind turbines (WTs) and erections in remote places, such as off-shore. The resulting increase of operation and maintenance costs can be counteracted by structural health monitoring systems. Different methods have been developed for detection of damages in WT blades. However, the majority are not suitable for in-service measurements or require dense sensor arrays. This paper presents a damage detection method based on autocorrelations of response accelerations. The damage sensitive feature (DSF) is developed as the Mahalanobis distance between a baseline and a current vector of the autocorrelation coefficients. Firstly, the usefulness of the DSF is assessed by using the Bayes error rate. Secondly, statistical hypothesis testing is utilized for a decision about the structural state. The procedure is applied to numerical simulations of a single WT blade with a disbonding damage scenario. The time series of accelerations are obtained from transient simulations with a simplified aerodynamic loading. The damage detection results show to be sensitive for the chosen damage scenario and are promising for prospective developments of damage detection methods in WTs.

11:30-11:50 TuAT6.4

Kalman-Filter Based Data Fusion for Neutral Axis Tracking for Damage Detection in Wind-Turbine Towers

Soman, Rohan — Inst. of Fluid Flow and Machinery, Pol. Acad. of Science
Malinowski, Pawel Henryk — Pol. Acad. of Sciences, Inst. of Fluid-Flow Machinery
Ostachowicz, Wiesław — Pol. Acad. of Sciences, Inst. of Fluid Flow Machinery

Wind Energy is seen as one of the most promising solutions to man's ever increasing demands of a clean source of energy. But there is a need to reduce the high initial costs for setting up and the maintenance costs. The maintenance cost may be lowered through the use of condition monitoring (CM) and structural health monitoring (SHM). SHM allows early detection of damage and allows maintenance planning which reduces the cost. In this paper, change in Neutral Axis (NA) position is proposed as a metric for damage detection. A discrete Kalman filter (KF) is employed for the estimation of the NA in the presence of measurement noise from the strain sensors. The KF allows data fusion from the strain sensors and the yaw mechanism for the accurate estimation of the NA. Any change in the NA position may be used as an indicator for the presence and location of the damage. The study has been carried out on the simulated FE model of the wind turbine tower and indicates that NA tracking based on data fusion is sensitive to damage and robust enough to overcome the effects of measurement noise, and yawing of the nacelle.

11:50-12:10 TuAT6.5

SHM Based System Design of a Wind Turbine Tower Using a Modal Sensitivity Based Bayes Detector

Hovgaard, Mads Knude — Ramboll, Aarhus Univ.
Brincker, Rune — Aarhus Univ.

It is investigated if material based structural safety can be replaced with safety obtained from SHM. SHM on its own does not add any value to a structure unless there is a decision policy attached. The further attachment of a loss function enables upfront calculation of the expected utility. In the presented case, damage sensitive features from sensitivity based damage detection are used in a supervised learning and optimization scheme. A Bayes detector, i.e. the static decision rule that minimizes the expected utility, is utilized for the system design. The technique is demonstrated in a simulation case of the NREL 5MW wind turbine tower subjected to bending fatigue and horizontal circumferential cracking at weld locations. Decision driven SHM is shown to change the initial design safety of the structure in the fatigue limit state. A common optimum of material safety and classifier threshold is found, enabling the calculation of the expected value of SHM. A sensitivity analysis of the feature extraction and of the loss function is included.

12:10-12:30 TuAT6.6

Effect of a Damage to Modal Parameters of a Wind Turbine Blade

Larsen, Gunner Chr. — Tech. Univ. of Denmark
Berring, Peter — Department of Wind Energy, Tech. Univ. of Denmark
Tcherniak, Dmitri — Bruel and Kjaer Sound and Vibration Measurement
Nielsen, Per Hørlyk — Tech. Univ. of Denmark
Branner, Kim — Department of Wind Energy, Tech. Univ. of Denmark

This study reports structural dynamic characteristics obtained experimentally from an extensive testing campaign on a 34m long wind turbine blade mounted on a test-rig under laboratory conditions. Further, these experimental results have been compared with analog numerical results obtained from a very detailed FE model of the same blade using 3D solid elements. Both an undamaged and a damaged blade are investigated, and it is observed that the natural frequencies of the first few modes of the blade change very little due to a significant artificial damage imposed in tailing edge, whereas the mode shapes - especially if decomposed into the flapwise, edgewise and torsional components - contain information which might be helpful for detecting and localizing wind turbine blade damages.

TuBT1 — Auditorium

Aeronautics II (Regular Session)

Chair: Paget, Christophe — Airbus
Co-Chair: Guemes, Alfredo — UPM

13:50-14:10 TuBT1.1

Evaluation of a UAV Composite Wing Spar Repair Using an Embedded Optical Fiber Rayleigh Back-Scattering Distributed Strain Sensing

Tur, Moshe — Tel-Aviv Univ.
Kressel, Iddo — Israel Aerospace Industries
Ben-Simon, Uri — Israel Aerospace Industries
Bergman, Arik — Tel-Aviv Univ.

Shemesh, N.N.Y. — IAF
Glam, Benny — IAF

A representing damaged UAV wing spar cap was repaired using bonded unidirectional composite materials. For such a repair concept, its strength and long-term durability depend on optimizing the geometry of the repair in order to minimize stress concentration in the adhesive. In order to assess the stress distribution along the repair bond-line, an optical fiber was embedded during the repair application for distributed Rayleigh back-scattering strain measurement. The fiber was placed inside the structure, therefore enabling taking strain measurements, at a high spatial resolution, under the wing skin directly on the spar repair. This type of sensing concept can also be used to monitor this critical repair over time. It also coincides with the recently introduced airworthiness requirements for UAVs, where the substantiation of structural bonded joint can be based on: "repeatable and reliable non-destructive inspection".

14:10-14:30 TuBT1.2

High Speed, In-Flight Structural Health Monitoring System for Medium Altitude Long Endurance Unmanned Air Vehicle

Kressel, Iddo — Israel Aerospace Industries
Osher, Shapira — IAF
Shemesh, N.N.Y. — IAF
Glam, Benny — IAF
Ariel, Dvorjeski — Israeli Air Force
Tur, Moshe — Tel-Aviv Univ.
Mashiach, Naama — IAF
Sovran, Ido — Tel-Aviv Univ.

This work presents the design, qualification and flight service evaluation of an embedded Fiber Bragg Gratings (FBG) based, Health and Usage Monitoring System (HUMS) for the Israeli Air Force Medium Altitude Long Endurance (MALE) Unmanned Aerial Vehicle (UAV). A total of 54 FBG sensors were embedded on the wing and tail booms, enabling accurate tracking of both the vibrations signature and the actual loading conditions of these components. Reliable flight data, obtained during flight tests and normal UAV operational missions were recorded and analyzed both in the frequency and time domains so that normal structural behavior could be identified and tracked. Based on the data obtained, it is now possible to track the loads enveloped during each flight, construct the landing impact vibration signatures and loading spectrum for the booms and wing. The sensing system has already gained over 1000 flight hours, including more than 300 logging hours. It is the purpose of this system to early detect and identify dangerous changes in the normal structural behavior of an individual UAV, prompting the required corrective action, thereby, paving the way to condition-based maintenance.

14:30-14:50 TuBT1.3

Flight Demonstration Testing with Distributed Optical Fiber Sensor

Saito, Nozomi — Mitsubishi Heavy Industries, LTD.

This paper reports flight demonstration test results using an airborne structural health monitoring system with distributed optical fiber sensor. The test was conducted using a business jet in order to verify operability and feasibility of the airborne monitoring system. An optical fiber sensor was attached on the front spar of the vertical tail and the measurement device was on the cabin floor of the business jet. Dynamic strain and temperature changes during flights were measured and those values were close to those measured by strain gages and a thermocouple. Static distributed strain and temperature changes during flights were also measured and those values denoted the same tendency to those measured by the strain gages and the thermocouple. Despite the good results, there were some errors on temperature measurement results. The authors, therefore, developed a new installation method to decrease the temperature measurement errors. These developments mean that the operability and feasibility of our monitoring system in actual flights were verified.

14:50-15:10 TuBT1.4

Aircraft Fastener Defect Detection and Characterization through Polar-Wavenumber Filtering of Full-Field Guided Wave Data

Haynes, Colin — Univ. of California, San Diego
Todd, Michael — Univ. of California, San Diego
Lee, Jung-Ryul — Chonbuk National Univ.

Scanning laser systems are a class of ultrasonic guided wave systems that have shown great promise for the inspection of complex structures because of the high spatial resolution of their resulting data. In this study, the focus is on applying full-field signal processing techniques to automatically detect and characterize fastener damage in aircraft components with realistic structural complexity. The particular testbed being considered is a composite tail section with several rivet lines. Artificial damage was introduced into several of the rivets with different magnitudes and orientations. While trained users of the laser scanning system can often detect the damage modes by simply viewing the resulting images, a more automated and quantifiable approach is desired to increase the consistency and quantify the confidence of the inspection system. In order to characterize the state of each rivet, a polar wavenumber processing technique is proposed. First, the full-field data is transformed to a new polar coordinate system centered on a particular rivet. A 3D Fourier transform then transforms the data to the frequency-wavenumber domain. Finally, the data is windowed to isolate the waves propagating in the positive radial direction. Analyzing the scattering pattern from each rivet in turn provides information on the health of each rivet. Damage detection may be accomplished either by comparison of the extracted scattering pattern with a model of the expected scattering pattern, or by an absolute reference-free method comparing rivets to one another to identify outliers. The results show that there is the potential to provide quantified damage detection and characterization for fastener damage using laser-excitation guided wave inspection systems.

15:10-15:30 TuBT1.5

An Online System for Remote NDT and SHM Operation with Content Adaptive Signal Compression

Westerkamp, Clemens — Univ. of Applied Sciences Osnabrueck
Hennewig, Alexander — Univ. of Applied Sciences Osnabrueck
Speckmann, Holger — Testia GmbH
Bisle, Wolfgang — Airbus Germany
Colin, Nicolas — EADS Innovation Works
Rafrafi, Mona — Testia France

Remote engineering systems are valuable tools to give visual assistance and remote support e.g. in NDT (Non-destructive Testing) or SHM (Structural Health Monitoring). They allow discussing a second opinion with a remote expert and thus reducing the human factor during testing and monitoring. For an optimal impression of the situation, the second person requires both a camera view of the location and the screen view of the system used. The OMA system (Online Maintenance Assistance) implements this two-view collaboration. Remote partners can see and actively control the equipment, while observing details of the location in the camera window. Due to varying working conditions, screen signals and communication properties, an adaptive compression for both signals (camera and screen) is proposed. This permits to always maintain the best possible visual quality for the assessment performed by the remote partner. The OMA screen compression is valuable for dynamic signals like in most NDT applications. Slower NDT and SHM applications benefit from a smoother and more realistic handling of the controlled software. Currently compression profiles are switched manually, but a classificator will soon allow automatic adaptation. Since most of the OMA system is browser based, it can be used on laptops or tablets and will be available for the iOS based iPad soon. The compression benefits can be used for sensor data compression and direct sensor data communication as well. New OMA integrations include other testing technologies, devices and other and other signal related aspects. The authors are interested in other collaboration scenario with the neeed for multiple views on cameras and screens.

15:30-15:50 TuBT1.6

Damage Location in Thick Composite Material Based on Acoustic Emission and Wireless Nodes

Paget, Christophe	Airbus Operations Limited
Patel, Roshan	Airbus Operations Limited

This investigation is studying the behaviour of a bespoke wireless impact locator design for monitoring structures against any kinds of impacts. The device performance was evaluated against its location repeatability, its ability to last the required service life, its power consumption and its wireless range. The main variables in this investigation were the sensor sensitivity, the wireless node data repeatability and the power supply output voltage. The investigation identified which power source to be used, what is the strategy to determine the remaining power life, as well as the feasibility to use such technology for impact location.

TuBT2 GH Room

Guided Waves II (Regular Session)

Chair: Abraham, Odile	IFSTTAR
Co-Chair: Dérobert, Xavier	IFSTTAR

13:50-14:10 TuBT2.1

A Study on the Coupling of Cylindrical Guided Waves in Thick-Walled Structures

Ziaja, Aleksandra	The Hong Kong Pol. Univ.
Cheng, Li	The Hong Kong Pol. Univ. Hung Hom, Kowloon, Hong Ko
Su, Zhongqing	The Hong Kong Pol. Univ.
Staszewski, Wieslaw J.	Department of Robotics and Mechatronics, AGH Univ. of Science and Technology, Kraków
Uhl, Tadeusz	AGH Univ. of Science and Technology, Kraków
Packo, Pawel	Department of Robotics and Mechatronics, AGH Univ. of Science and Technology, Kraków

In this study, systematic analysis of the wave propagation characteristics in a thick-walled cylindrical structure was achieved with a purpose of facilitating the development of a damage detection system for train axles. The correspondence of thick hollow cylinder waves to the plate wave solutions of symmetric and antisymmetric modes, and the influence of thickness and radius on wave propagation were examined through the asymptotic approximations of Bessel functions. The analytical solutions to dispersion characteristics were compared with numerical simulations using the Local Interaction Simulation Approach and experimental validation.

14:10-14:30 TuBT2.2

Mode Selective Actuator-Sensor System for Lamb Wave-Based Structural Health Monitoring

Schmidt, Daniel	German Aerospace Center (DLR), Inst. of Composite Structures
Wierach, Peter	German Aerospace Center (DLR), Inst. of Composite Structures
Sinapius, Michael	Tech. Univ. Braunschweig, Inst. of Adaptronics an

Structural Health Monitoring (SHM) based on Lamb waves, a type of ultrasonic guided waves, is a promising method for in-service inspection of composite structures. In this study mode selective actuators and sensors are investigated to excite a particular Lamb wave mode in composite plates. The actuator and sensor exhibit an interdigital transducer design. In order to describe the complex displacement fields of A_0 and S_0 mode and to characterize the mode selectivity of the transducers a two dimensional analytical model based on higher order laminated plate theory is developed.

14:30-14:50 TuBT2.3

Long Term Monitoring of an Aircraft Structure During a Full Scale Fatigue Test

Dziendzikowski, Michal	Air Force Inst. of Tech.
Dragan, Krzysztof	Air Force Inst. of Tech.
Artur, Kurnyta	Air Force Inst. of Tech.
Klysz, Sylwester	Air Force Inst. of Tech.

In the paper a technique for qualitative assessment of fatigue crack growth monitoring is presented, utilizing guided elastic waves generated by sparse PZT piezoelectric transducers network in the pitch catch configuration. The Damage Indices used for the inference carries marginal signal information content in order to decrease their sensitivity with respect to undesired non-controllable factors. The reason for that is to limit the false calls ratio which besides the damage detection capability of a system, plays a crucial role in applications. However even such simplified damage indices can be altered over a long term, leading to the misclassification problem. Considering single sensing path, it is very difficult to distinguish whether the resultant change of DIs is caused by a damage or due to such DIs decoherence. Therefore assessment approaches based on threshold levels fixed separately for DIs obtained on each of the sensing paths, would eventually lead to a false call. In order to decrease such misclassification risk a method to compensate the DIs drift is proposed utilizing the information from all of the network sensing paths. The proposed approach has been verified on a real structure during a Full Scale Fatigue Test (FSFT).

14:50-15:10 TuBT2.4

Damage Detection in Composites by Noncontact Laser Ultrasonic,

Park, Byeongjin	KAIST
Sohn, Hoon	Professor, Department of Civil and Environmental Engineering, KA
Malinowski, Pawel Henryk	Pol. Acad. of Sciences, Inst. of Fluid-Flow Machinery
Ostachowicz, Wiesław	Pol. Acad. of Sciences, Inst. of Fluid Flow Machinery

This study proposes an instantaneous damage localization technique for composite structures using noncontact laser ultrasonics. First, a velocity profile of the target composite structure is obtained by measuring wave velocities on paths with various directions. Second, laser ultrasonic responses from two excitation-sensing pairs, called inspection pairs, are obtained. Then possible damage locations are estimated through time-of-flight triangulation of damage reflected waves. Once possible damage locations are estimated, ultrasonic response collection is repeated near the estimated locations for precise damage localization. The proposed technique has following advantages over existing technique: (1) It does not require any sensor installation; (2) Wave velocity profile, which has been mandatory information for time-of-flight triangulation, can be easily experimentally constructed; and (3) It can localize damage with a high precision with its scanning capability. The feasibility of the proposed technique is examined by localizing a delamination in a real 10 kW wind turbine blade.

15:10-15:30 TuBT2.5

A Pattern Recognition Approach for Damage Detection and Temperature Compensation in Acousto-Ultrasonics

Torres Arredondo, Miguel Angel	MAN Diesel & Turbo SE
Sierra, Julian	Univ. Pol. de Madrid
Zenuni, Erion	Univ. of Siegen
Cabanes, Guenael	Paris XIII Univ.
Rodellar, Jose	Tech. Univ. of Catalonia
Guemes, Alfredo	UPM
Fritzen, Claus-Peter	Uni Siegen

The global trends in the construction of modern structures require the integration of sensors together with data recording and analysis modules so that their integrity can be continuously monitored for safe-life, economic and ecological reasons. This process of measuring and analysing the data from a distributed sensor network all over a structural system in order to quantify its condition is known as structural health monitoring (SHM). Guided ultrasonic wave-based techniques are increasingly being adapted and used in several SHM systems which benefit from built-in transduction, large inspection ranges, and high sensitivity to small flaws. However, for reliable health monitoring, much information regarding the innate characteristics of the sources and their propagation is essential. Moreover, any SHM system which is expected to transition to field operation must take into account the influence of environmental and operational changes which cause modifications in the stiffness and damping of the structure and consequently modify its dynamic behaviour. On that account, special attention is paid in this paper to the development of an efficient SHM methodology where robust signal processing and pattern recognition techniques are integrated for the correct interpretation of complex ultrasonic waves within the context of damage detection and identification. The methodology is based on an acousto-ultrasonics technique where the discrete wavelet transform is evaluated for feature extraction and selection, linear principal component analysis for data-driven modelling and self-organizing maps for a two-level clustering under the principle of local density. At the end, the methodology is experimentally demonstrated and results show that all the damages were detectable and id

TuBT3	Room I
Environmental Effects (Regular Session)	
Chair: Kullaa, Jyrki	Helsinki Metropolia Univ. of Applied Sciences
Co-Chair: Theroux, Louis-Daniel	IFSTTAR

13:50-14:10 TuBT3.1

Compensation of Temperature Effects on Guided Wave Based Structural Health Monitoring Systems

Dan, Codrut Alexandru — Inst. of Fluid Flow Machinery, Pol. Acad. of Sciences
Kudela, Pawel — Pol. Acad. of Sciences, IFFM
Ostachowicz, Wiesław — Pol. Acad. of Sciences, Inst. of Fluid Flow Machinery

The study focuses on the development of analytical methods to compensate for the temperature effects on guided wave propagation. The physical effects of temperature upon the registered signals using piezoelectric sensors mounted in pitch-catch configuration are explored experimentally over considerably large temperature range and frequency bandwidths. The observations made are mathematically interpreted and a mathematical framework which enables a new strategy for compensation of the temperature effects on Lamb waves is formulated. The strategy is validated in an experimental campaign where the plate with the installed SMART Layer is monitored for low temperature differences, at room conditions, over longer periods of time. Damage scenarios under the form of added mass were considered and the influence of temperature on the damage detection and the efficiency of the proposed strategy are investigated.

KEYWORDS : Structural Health Monitoring (SHM), Guided Waves, Piezoelectric Sensors, Temperature Effects, Signal Processing.

14:10-14:30 TuBT3.2

OFDR Distributed Temperature and Strain Measurements with Optical Fibre Sensing Cables: Application to Drain Pipeline Monitoring in a Nuclear Power Plant

Maurin, Laurent — CEA, LIST, Optical Measurement Lab. F-91191 Gif-sur-Yvette
Rougeault, Stéphane — CEA
Dewynter-Marty, Véronique — CEA
Périsse, Jocelyn — AREVA
Villani, Dominique — NNS
Macé, Jean-Reynald — AREVA
Ferdinand, Pierre — CEA LIST

This study deals with the testing of innovative Optical Fibre Sensing (OFS) cables deployed on ducts, with the aim to perform distributed temperature and strain measurements. Such cables contain several optical fibres devoted to be interrogated by Optical Frequency Domain Reflectometry (OFDR). The methodology has first been developed and qualified in laboratory. Then, real tests have been performed on a Nuclear Power Plant (NPP) drain system to demonstrate the industrial feasibility of such technology. To do so, two small diameter sensing cables, compatible with distributed temperature and strain measurements, have been qualified and afterwards installed along a sodium drain line at Superphénix NPP (liquid sodium coolant fast breeder reactor in current dismantling). Measurements have been performed during the preheating operation. Recorded data were post-processed according to a semi-empirical model taking into account temperature dependence and thermo-mechanical sensing cable behaviour. Optical fibre distributed temperature measurements were then successfully compared to thermocouple reference measurements, whereas optical sensing cable data were processed to provide distributed strain, then distributed curvature radius, which will enable, after numerical integration, to compute distributed displacement data. The goal is to assess the use of OFS for monitoring both temperature and mechanical strain distribution along a pipe under heat stress.

14:30-14:50 TuBT3.3

ADALINE Network-Based Temperature Compensation Method for SHM Method

Huang, Tianxiang — Nanjing Univ. of Aeronautics and Astronautics
Yuan, Shenfang — Nanjing Univ. of Aeronautics and Astronautics
Qiu, Lei — Nanjing Univ. of Aeronautics and Astronautics

The performance of current PZTs array and Lamb wave based SHM methods is restricted by the temperature variation. As temperature has a significant influence on Lamb wave propagation which leading to false monitoring result even if the structure is under health status. A new temperature compensation method based on adaptive filter adaptive liner neural (ADALINE) network has been developed to compensate for the amplitude-change and phase-shift of Lamb wave due to temperature change. Network construction and procedure of compensation are discussed. The advantages of this proposed method are (1) Simple to be applied, (2) only a few baselines are required for a large temperature range. Experiments are conducted on a stiffened carbon fiber composite panel to verify the temperature compensation method under temperature range from -40°C to 80°C. Results show the presented method is effective. Damage image combined with this compensation method can detect the damage under temperature variations.

14:50-15:10 TuBT3.4

Analysis of Environmental and Operational Condition Effects on Guided Ultrasonic Waves in Stiffened CFRP Structures

Eckstein, Benjamin — EADS Innovation Works
Moix-Bonet, Maria — DLR
Bach, Martin — EADS Deutschland GmbH

Structural Health Monitoring by Guided Ultrasonic Waves for real world applications relies on compensation of operational and environmental effects. The effects to be taken into account are depending not only on the present factors for a particular application scenario, but as well on the structure to be monitored itself. Especially for stiffened CFRP structures, wave propagation and subsequent temperature effects are of increasing complex nature. The load and temperature condition of aerospace components typically vary along the flight cycle of an aircraft and is subject of a vast number of operational and environmental factors. The extensive research on temperature influence on wave propagation and transduction as the most important influence factor in the last years has broad up two widely spread methods for temperature compensation: Optimum Baseline Selection and Baseline Signal Stretch, which can be applied in a combination as well. In this work the existing compensation concept are challenged regarding its applicability to real world structures and environmental and operational conditions. For that purpose experimental data from two structures are analysed. Firstly, the influence of on wave propagation is analysed in the scope of a lab test by data from a stiffened CFRP panel subject to temperature variation. Secondly, the influence of environmental and operational conditions in the frame of a flight test is analysed.

15:10-15:30 TuBT3.5

Validation of Temperature Compensation Techniques for Impact Damage Detection and Localization Using Ultrasonic Sparse Arrays

Scheerer, Michael — Aerospace & Advanced Composites GmbH
Lager, Daniel — AIT Austrian Inst. of Tech. GmbH

Within this paper the authors implemented and tested three different methods for temperature compensation – best baseline approach, signal interpolation and frequency shift – on thin monolithic composite plates and composite honeycomb panels together with the effectiveness of the different approaches for damage detection and localization under varying temperature conditions using sparse array configurations. The effectiveness of the different procedures have been compared and ranked. The best suited procedure – the frequency shift approach using the closest baseline measurement – was implemented in virtual beam forming algorithm used to visualize the ultrasonic image of the structure. Experiments have been conducted on an 800 x 800 mm² honeycomb panel instrumented with 4 piezo actuators and 8 piezo sensors. Different baseline measurements at various temperatures have been compared against measurement after the application of artificial damages using the best suited temperature compensation technique. It could be shown that temperature differences of several degrees between the baseline measurement and the measurement after damage introduction completely hides the appearance of the damage without using temperature compensation techniques. On the other hand all damages could be successfully detected and located when applying the proposed temperature compensation technique.

TuBT4 Room J
Bridge Monitoring (Regular Session)

Chair: Akhras, George — Centre for Smart Materials and Structures - Royal Military Coll. of Canada
Co-Chair: Hille, Falk — BAM Federal Inst. for Materials Res. and Testing

13:50-14:10 TuBT4.1

Modelling the Human Induced Vibrations in a Cable-Stayed Pedestrian Timber Bridge

Casciati, Sara — Univ. of Catania
Casciati, Fabio — Univ. of Pavia
Faravelli, Lucia — Univ. of Pavia
Bortoluzzi, Daniele — Univ. of Pavia

In this paper the data acquired on a pedestrian timber bridge under different "moving" loads configurations are reported and analysed by applying a time-frequency decomposition technique. This approach allows the authors to better identify the dynamic behavior of the bridge under the above loads. A first in situ experimental campaign was carried out on November 7, 2013, with the aim of recording the accelerations induced by people walking and running along the bridge (the so called human induced vibration – HIV). For this purpose, the footbridge was equipped by accelerometers. The pedestrian bridge under investigation is located not far from the town of Belluno, Italy, and connects the two sides of the outlet channel of the "Santa Croce" Lake. To better fit the structure to the surrounding naturalistic area, eco-friendly construction materials with low environmental impact were adopted during the design stage. In particular glued laminated timber (GLT) and steel elements are used, limiting the use of reinforced concrete only for the foundation system.

14:10-14:30 TuBT4.2

Integrated System for Continuous Dynamic Monitoring: "Ponte Nuovo Del Popolo" Bridge

Caldon, Mauro — Univ. of Padova
Islami, Kleidi — Univ. of Padova
Lorenzoni, Filippo — Univ. of Padova
Modena, Claudio — Univ. of Padova

The Italian territory is characterized by a high seismic risk and the presence of several existing structures and infrastructures leads to the definition of accurate procedures and methodologies to evaluate their safety conditions and the life-cycle performance. In this framework Structural Health Monitoring (SHM) is currently more and more considered as a key activity to increase the knowledge on the structural behavior of existing bridges. The paper reports the on-site diagnostic investigations and the installation activities of a SHM system of the "Ponte Nuovo del Popolo" bridge, located in the city center of Verona. The results of the first 1.5 year of monitoring are presented, analysing the operational conditions of the structure subjected to both static and dynamic loads. Specific algorithms and processing software were developed and implemented to perform the continuous real time treatment and analysis of static data and the automated extraction of modal parameters based on the measurement of the bridge's response to ambient vibrations. The bridge is subjected to widespread damaging phenomena, mainly concentrated on the external longitudinal beams. The SHM system is composed by six uniaxial accelerometers and 16 linear displacement transducers and its acquired data are constantly related to the environmental parameters (temperature and relative humidity).

14:30-14:50 TuBT4.3

Case Studies Illustrating the Effect of Weakened Load Transfer on Neutral Axis Measurements

Gangone, Michael — Univ. of Texas at Tyler

Experimental load rating, utilizing strain measurements obtained under controlled vehicle loading, allows for the structures capacity to be determined. This is particularly important for older bridges. However, deterioration can often hinder the load transferring capabilities of the bridge. This results in weaker signal strength at girders farthest from the applied loading. If this is not recognized the results from the load testing and subsequent load rating may prove erroneous. This paper discusses the accuracy of one particular parameter, the neutral axis of bending based on the ability of the structure to adequately distribute the load. Two separate multi-girder/stringer bridges at different periods of their service life were tested using a custom developed wireless sensor system under various loading scenarios. The neutral axis and load distribution factors were acquired using standard department of transportation practices. The results show the ineffectiveness of the deteriorated bridge to transfer the load to resulting in highly variable measures of the neutral axis at girders farthest from the loading. However, this variability was minimal in the newer bridge.

14:50-15:10 TuBT4.4

Innovative Investigations in a Prestressed Concrete Bridge

Houel, Adrien — CEREMA, DTCE
Valade, Michel — CEREMA, DTCE, DLL
Germain, Didier — CEREMA, DTCE, DLL
Mathey, Auguste — CEREMA, DTCE, DLL

In France most of prestressed concrete box-girder bridges are the object of reinforcements these last years per addition of prestressing tendons in most cases following classical complementary investigations carried out within the framework of project of repair. An innovating investigation was realized on a motorway viaduct close to Lyon in May 2013: strain gauges will be positioned on four steel tendons of prestressed right to significant cracks of the box-girder, cracks also instrumented by displacement sensors. Indeed, risks of overstretching in the tendons are expected under road traffic. A first relation will be established between the overstretchings evaluated from the measurements of the gauges and the openings of cracks from the dis-placement sensors. Moreover, measurements of temperature gradients were also carried out to distinguish solicitations from traffic loads and temperature effects. All these measurements are realized from a new Wi-Fi platform called PEGASE developed by IFSTTAR and the society A3IP. Finally, a fatigue analysis has been carried out in order to establish the residual lifetime of the bridge. An analysis in robustness will be undertaken. Three approaches are described in this paper. The most realistic approach is the experimented one. At the conclusion of this analysis, a decision will be made by the owner based on the opinion of experts to possibly launch the project of heavy repair of the bridge per external additional prestressing.

15:10-15:30 TuBT4.5

Damage Detection Based on Structural Response to Temperature Changes and Model Updating

Ralbovsky, Marian	AIT Austrian Inst. of Tech. GmbH
Santos, João	Lab. Nacional de Engenharia Civil (LNEC)
Kwapisz, Maciej	AIT Austrian Inst. of Tech.
Dallinger, Sonja	RED Bernard
Catarino, Jose Manuel	Lab. Nacional de Engenharia Civil (LNEC)

The paper proposes use of measured structural response to temperature loads for purposes of damage identification. As opposed to the most common approaches, which rely on suppressing temperature effects in damage detection, the method proposed herein utilizes measured changes of structural sensitivity to temperature loads. Applicability of this approach varies by structural type and therefore results of system identification are presented for different structural types. The method aims at identifying structural damages that produce stiffness change as well as deformation states and boundary condition stiffness. In particular, the method is applicable for evaluations without available measurements known as healthy baseline. Presented examples comprise 2 bridge monitoring applications.

15:30-15:50 TuBT4.6

Inferring Asset Live Load Distributions from Traffic Flow Data: A New SHM Opportunity?

Sousa, Helder	Univ. of Surrey
Zavitsas, Konstantinos	Imperial Coll. London
Polak, John	Imperial Coll. London
Chryssanthopoulos, Marios	Univ. of surrey

With the continuous ageing of infrastructures, live load distributions related to actual traffic has become one of the key inputs in asset management. However, it is also one fraught with difficulty, due to its complex and dynamic nature, which can only be addressed at a network level. At this level, it is impossible to envisage in-situ SHM systems installed for all critical assets. In this context, the development of reliable alternative techniques to estimate live load distributions would be a valuable addition to infrastructure asset management tools. As is well known, the accuracy of such estimates depends on several factors such as road capacity, asset condition/performance, traffic composition and seasonal effects, among others. Thus, the deployment of in-situ asset-specific systems needs to be complemented with other types of monitoring systems based on inexpensive and easy to install traffic flow sensors (point and line) in order to infer, with acceptable accuracy, rather than measure directly the live load distributions pertinent to different asset types on the network. This paper presents an approach to derive load distributions based on a Transport Infrastructure Utilisation and Maintenance Framework by utilizing recent advances achieved in two, often non-communicating, fields: structural engineering and transportation engineering. From the realm of transport analysis, the parameter 'flow' has been combined with the parameter 'live load' pertinent to structural performance. Taking advantage of traffic flow sensor systems, the aim is to examine how information related to the former enables the understanding and modelling of the latter, thus paving the way for smart transport mobility technology to be harnessed by the structural asset management community.

TuBT5 KL Room

Sensors II (Regular Session)

Chair: Pons, Patrick	LAAS-CNRS
Co-Chair: Cumunel, Gwendal	Lab. Navier (ENPC / IFSTTAR / CNRS)

13:50-14:10 TuBT5.1

Development of High-Performance BS-PT Based Piezoelectric Transducers for High-Temperature SHM Applications

Li, Yu-Hung	Stanford Univ.
Kim, Sang-Jong	Stanford Univ.
Salowitz, Nathan	Stanford Univ.
Chang, Fu-Kuo	Stanford Univ.

This paper focuses on developing new BiScO3-PbTiO3 (BS-PT) based piezoelectric ceramic transducers for high-temperature SHM applications. By controlling the PbO deficiency in the material system, we modify the lossy ferroelectric properties and enhance the piezoelectric responses from the intrinsic BS-PT. The new in-house fabricated piezoelectric transducers can maintain comparably high piezoelectric responses at temperatures up to 350°C continuously for at least 10 hours. The signals of the new BS-PT based piezoelectric transducers were tested and studied at different temperatures on metallic structures, and we observed a promising trend of obtaining stronger signals from the BS-PT transducers at higher temperatures. A new microfabrication method was developed to miniaturize and to mass-fabricate the piezoelectric transducers for potential applications on complex structures in extreme environments.

14:10-14:30 TuBT5.2

Handy Microwave Sensor for No-Contact Detection of Structural Vibration

Pieraccini, Massimiliano	Univ. di Firenze
Donati, Nicoletta	Univ. degli Studi di Firenze
Papi, Federico	Univ. degli Studi di Firenze

The authors propose a novel handy microwave sensor for monitoring structures as bridges, towers, streetlight, floors. It operated at distance by detecting the natural frequency of the structure under test. The device has been designed as a handy instrument with a friend user-interface and a simple measurement procedure. The aim is to provide a portable equipment for the engineering studies. The prototype has been tested in laboratory conditions and in a preliminary in-field measurement session on a streetlight.

14:30-14:50 TuBT5.3

Voice-Coils As Reciprocal Transducers in Structural Health Monitoring Applications

Fröjd, Patrik	Lund Univ.
Ulriksen, Carl Peter Fredrik	Lund Univ. Faculty of Engineering

The feasibility of using voice-coils as reciprocal transducers of mechanical waves is investigated. It is speculated that these transducers could be used in a seismic network for structural health monitoring purposes, transmitting relatively un-damped low frequency waveforms such as surface waves. This would permit long distances between transducers. The force generation capacity and vibration sensitivity of five types of voice-coil transducers are compared in the lower sonic spectrum. As a verification of the concept, the transducer type evaluated to be the most practical is used to measure the natural bending mode frequencies of a small concrete beam.

14:50-15:10 TuBT5.4

Piezoelectric Paint Sensor for Impact and Vibration Monitoring

Kang, Lae-Hyong	Chonbuk National Univ.

This paper presents a fabrication method for a piezoelectric paint sensor, and its application to impact and vibration monitoring of a beam structure. The piezoelectric paint in the paper is composed of Pb(Nb,Ni)O3-Pb(Zr,Ti)O3 (PNN-PZT) powder and epoxy resin. The piezoelectric paint has been coated on an aluminum beam structure, and the electrode has been made on the upper surface of the piezoelectric paint using silver paste. By using the aluminum structure as the opposite electrode, the poling of the PNN-PZT/Epoxy paint has been conducted successfully at room temperature with the electric field of 4 kV/mm for 30 min. When the cantilevered aluminum beam was under several vibration conditions, especially resonance, the vibration response has been monitored by measuring the output voltage from the piezoelectric paint sensor. The output voltage responses to the in-plane deformation(strain) of the beam structure, so the displacement-strain transformation technique has been used in order to obtain out-of-plane deformation. The transformed out-of-plane deformation from the paint sensor was compared with the measured from laser displacement sensors at a certain beam position. The results showed that the estimated data matches the measured and, moreover, the deformations at various points along the beam length also can be obtained from the displacement-strain transformation technique. In addition to the vibration, impact signal also can be easily monitored using the piezoelectric paint, because the piezoelectric paint is inherently more sensitive to higher frequency signals. By using the impact hammer, several impacts have been applied to the beam structure. The impact monitoring test results showed that the impact force signal over 10N can be captured using the piezo-paint.

15:10-15:30 TuBT5.5

Tunable Interdigital Transducer for Lamb Waves

Manka, Michal	AGH Univ. of Science and Technology, Kraków
Rosiek, Mateusz	AGH Univ. of Science and Technology, Kraków
Martowicz, Adam	AGH Univ. of Science and Technology, Kraków
Stepinski, Tadeusz	AGH Univ. of Science and Technology, Kraków
Uhl, Tadeusz	AGH Univ. of Science and Technology, Kraków

In this paper a novel concept of a tunable transducer for exciting and sensing Lamb waves is presented. The proposed transducer is an extension of the Interdigital Transducer (IDT), which main advantage is mode selectivity. In the proposed transducer, further referred to as the Tunable IDT (T-IDT), comb electrodes with fixed pitch are replaced by a series of densely distributed discrete electrodes. They can be easily interconnected in different configurations to match required wavelength, which makes possible matching to different wavelengths without a need of physical changes of the electrodes' layout. In order to verify the properties of the T-IDT, numerical simulations and experimental tests were performed and their results compared to the ones obtained with a traditional IDT. Both the numerical and experimental results show that the T-IDT generates beampatterns similar to those generated by the conventional IDT. The preliminary results presented in the paper prove that the proposed T-IDT can be easily tuned to various wave modes and show the performance similar to the one observed for the conventional IDT.

15:30-15:50 TuBT5.6

Calibration Problem of AD5933 Device for Electromechanical Impedance Measurements

Wandowski, Tomasz	Inst. of Fluid-Flow Machinery PAS, 80-231 Gdansk, Pol.
Malinowski, Pawel Henryk	Pol. Acad. of Sciences, Inst. of Fluid-Flow Machinery
Ostachowicz, Wiesław	Pol. Acad. of Sciences, Inst. of Fluid Flow Machinery

In this paper influence of incorrect calibration of AD5933 electrical impedance analyzer on errors in measurements was investigated. AD5933 is developed in the form of electronic integrated circuit by the Analog Devices. This device is very popular in the field of Structural Health Monitoring based on electromechanical impedance method. Electromechanical impedance method relies on measurements of electrical impedance of piezoelectric transducer bonded to the structure. Due to electromechanical coupling of piezoelectric transducer mechanical resonances of the structure can be seen in the electrical impedance characteristics of piezoelectric transducer. In order to properly measure electrical impedance mentioned devices must be calibrated using calibration resistor. In this research influence of wrongly selected calibration resistor value on the measurement results was interrogated. Measurements conducted with AD5933 were compared with measurements realized using professional electrical impedance analyzer HIOKI IM3570. Research was related to simple supported beam made out of aluminium alloy with piezoelectric transducer in the form of disc. Measurements were taken for referential state as well as for few states with artificial introduced damage in the form of notch.

TuBT6 Room 200
Wind Energy II (Regular Session)

Chair: Fritzen, Claus-Peter	Uni Siegen
Co-Chair: Flynn, Eric	Los Alamos National Lab.

13:50-14:10 TuBT6.1

Detection of Impulse-Like Airborne Sound for Damage Identification in Rotor Blades of Wind Turbines

Krause, Thomas	Leibniz Univ. Hannover
Preihs, Stephan	Univ. Hannover
Ostermann, Jörn	Leibniz Univ. Hannover

Structural health monitoring systems can help to improve safety and minimize the numerous economical burdens of wind turbines. To detect damage of rotor blades, several research projects focus on an acoustic emission approach. Acoustic emission stands for stress waves emitted by a damage process. For this approach components of the waves are measured with sensors mounted on the surface of the blade. Small damages can be detected, but the amount of sensors is relatively high due to the size of modern blades and high internal damping of composite materials. The damage process and stress waves also emit airborne sound. In contrast to existing approaches we use the airborne sound for damage detection. We developed a detection algorithm based on our signal analysis which showed that airborne sound provides adequate features for cracking sound detection. We optimized and tested the algorithm with our airborne sound recordings of a full-scale rotor blade test. In the first three days of the long term fatigue test our algorithm detects nine events per day, 79% of the events are cracking sounds. In the remaining 73 days of the fatigue test the algorithm detects about one event per day.

14:10-14:30 TuBT6.2

Use of a Spatially Adaptive Thresholding Method for the Condition Monitoring of a Wind Turbine Gearbox

Antoniadou, Ifigeneia	The Univ. of Sheffield
Worden, Keith	Univ. of Sheffield

Condition monitoring of wind turbine gearboxes is an important practice in order to determine the state of the wind turbine drivetrain. In this way reparative actions could be taken whenever needed, resulting in reduction of maintenance costs. In this paper, time-frequency analysis is performed on real wind turbine gearbox datasets using the empirical mode decomposition method. Then, the outlier analysis method is applied to the power of certain intrinsic mode functions of the decomposed - using the empirical mode decomposition method - gearbox experimental datasets. These intrinsic mode functions are chosen according to their frequency content. They are related to the harmonics of the meshing frequency of the damaged stage of the gearbox examined. The outlier analysis method is a well- established method in the structural health monitoring field that computes discordancy measures for data and compares them with a threshold. Here, it is used as a standard approach whose results can be used for comparison. Finally, a novel thresholding method is proposed for feature discrimination - the phase space thresholding method. It is shown that for the particular case of gear tooth damage, because of the way it manifests in the vibration signals, the phase space thresholding method proves to be a very satisfactory method that can be used for an enhanced condition monitoring strategy.

14:30-14:50	TuBT6.3

Fault Detection and State Evaluation of Rotor Blades

Petryna, Yuri	Tech. Univ. Berlin
Künzel, Andreas	Tech. Univ. Berlin
Kannenberg, Matthias	Tech. Univ. Berlin

The present work has been done within the Joint Research Project Blade Tester, which develops an approach for cost-effective, serial integrity tests of rotor blades. At that, manufacturing faults shall be automatically detected, localized and assessed with respect to their impact on the integrity of rotor blades in operation. Several techniques are involved and discussed in the contribution, such as 3D full-scale geometry measurement, structural modelling by the finite element method and model validation, fibre optical sensing, several non-destructive testing techniques and parameter identification on the basis of static and dynamic tests.

14:50-15:10	TuBT6.4

Wind Turbine Structural Health Monitoring: A Short Investigation Based on SCADA Data

Papatheou, Evangelos	Univ. of Sheffield
Dervilis, Nikolaos	Univ. of Sheffield
Maguire, Eoghan	Vattenfall Res. & Development
Worden, Keith	Univ. of Sheffield

The use of offshore wind farms has been growing in recent years, as steadier and higher wind speeds can be generally found over water compared to land. Moreover, as human activities tend to complicate the construction of land wind farms, offshore locations, which can be found more easily near densely populated areas, can be seen as an attractive choice. However, the cost of an offshore wind farm is relatively high, and therefore their reliability is crucial if they ever need to be fully integrated into the energy arena. As wind turbines have become more complex, efficient, and expensive structures, they require more sophisticated monitoring systems, especially in offshore sites where the financial losses due to failure could be substantial. This paper presents the preliminary analysis of supervisor control and data acquisition (SCADA) extracts from the Lillgrund wind farm for the purposes of structural health monitoring. A machine learning approach is applied in order to produce individual power curves, and then predict measurements of the power produced of each wind turbine from the measurements of the other wind turbines in the farm. A comparison between neural network and Gaussian process regression is also made.

15:10-15:30	TuBT6.5

Localization of Defects in Pipes Using Guided Waves and Synthetic Aperture Focussing Technique (SAFT)

Schubert, Lars	Fraunhofer
Weihnacht, Bianca	Fraunhofer IKTS-CMD Dresden
Gaul, Tobias	Fraunhofer IKTS-CMD Dresden
Frankenstein, Bernd	Fraunhofer IKTS-CMD Dresden

For maritime applications like offshore foundations of wind turbines, durable sensors and sensor systems with low noise levels are needed which are well suited for the harsh environmental conditions. Additionally, the restricted offshore repair opportunities and therefore the urgent need for high reliability have to be taken into account. In order to monitor welded seams in these applications areas, sensor rings were tested and qualified. Sensors are positioned in a ring around the pipe. The area of the welded seam is investigated by ultrasonic waves. Especially the correct positioning of the damage demands for highly sophisticated processing algorithms which are able to reconstruct the damage reliably from the recorded signals. The present paper introduces a technique to localize cracks in cylindrical structures. The proposed technique is based on the use of elastic waves propagating in hollow cylinders The so-called guided waves are used in a variety of wave modes. Every wave mode induces a different interaction potential with a crack depending on frequency and elastic stress components. Using the Finite difference method, the time-dependent elastic problem is solved by the simulation software Wave3000Plus to determine well-suited wave modes and frequency domains for the interaction of waves with the expected cracks. Subsequently, measurements using a cylindrical test structure were performed to verify the simulation results. Varying wave modes provide differing main oscillation directions and require the usage of shear transducers and thickness crystal transducers. Furthermore, the synthetic superposition technique is approved by experiments on cracks with varying depth.

TuCT1	Auditorium
Composites I (Regular Session)	
Chair: Giordano, Michele	CNR-IMCB Inst. for Composite and Biomedical Materials
Co-Chair: Drissi-Habti, Monssef	IFSTTAR

16:10-16:30	TuCT1.1

Damage Detection in Composite Structures from Fibre Optic Distributed Strain Measurements

Guemes, Alfredo	UPM
Fernandez-Lopez, Antonio	UPM

Two approaches may be followed to obtain damage information from strain measurements: 1) To use multiple sensing points, and to reduce the huge amount of information by pattern recognition techniques, as PCA. (companion paper) 2) To use distributed sensing (OFDR of backscattered radiation), placing the plain optical fibre at the region where damage is expected. Because of its simplicity, this approach is very robust, and even its local nature, similar to the VCM, it may be of interest in some applications, like stringer debonding, or for identifying local damage by impacts around man-holes. One advantage of this approach compared to the former one, is that the structure does not need to be under load, it measures the residual strains caused by the damage. Several experiments, done on real aeronautic composite parts, are presented, quantifiying the POD, robustness and applicability of the technique.

16:30-16:50	TuCT1.2

A Heat Flux Sensor to Monitor Composite Manufacturing Processes

Villiere, Maxime	Lab. de Thermocinétique de Nantes UMR CNRS 6607
Tardif, Xavier	IRT Jules Verne
Gueroult, Sebastien	IRT Jules Verne
Sobotka, Vincent	Univ. of Nantes, CNRS, Lab. de Thermocinétique
Boyard, Nicolas	CNRS/Lab. de Thermocinétique de Nantes
Delaunay, Didier	CNRS, Lab. de Thermocinétique, Pol.
Breard, Joël	Univ. of Le Havre

Composite materials are heterogeneous materials which consist of a matrix reinforced by fibers. One of the most promising manufacturing processes are Liquid Composite Moulding (LCM) in which a stationary fibrous preform is impregnated by a liquid. It is essential to monitor two key-steps in order to ensure high quality composite parts: the mould-filling stage, during which the liquid resin flows through the fibrous preform and the solidification. Two examples related to these issues are detailed here, the common point being the use of the same sensor to identify the different properties. In a first part, an experimental device has been designed to quantify the saturation profile of a preform. This set-up consists of a mould with heat flux sensors and an injection device reproducing the actual forming process. The saturation curve can be obtained by exploiting the wall heat fluxes. In a second part, this study emphasizes the use of the same heat flux sensors to monitor the consolidation phase. Indeed heat transfer and temperature during the forming process are strongly coupled with the phase change through the variation of thermal properties. These methods clearly open up new perspectives regarding the online control of composite manufacturing processes at an industrial scale.

16:50-17:10 TuCT1.3

Vibro-Acoustic Modulation Based Damage Identification in a Composite Skin-Stiffener Structure

Ooijevaar, Ted — Univ. of Twente
Loendersloot, Richard — Univ. of Twente
Rogge, Matt — NASA Langley Res. Center
Akkerman, Remko — Univ. of Twente
Tinga, Tiedo — Univ. of Twente

The vibro-acoustic modulation method is applied to a composite skin-stiffener structure to investigate the possibilities to utilise this method for damage identification in terms of detection, localisation and damage quantification. The research comprises a theoretical part and an experimental part. An impact load is applied to the skin-stiffener structure, resulting in a delamination underneath the stiffener. The structure is interrogated with a low frequency pump excitation and a high frequency carrier excitation. The analysis of the response in a frequency band around the carrier frequency is employed to assess the damage identification capabilities and to gain a better understanding of the modulations occurring and the underlying physical phenomena. Though vibro-acoustic is shown to be a sensitive method for damage identification, the complexity of the damage, combined with a high modal density, complicate the understanding of the relation between the physical phenomena and the modulations occurring.

17:10-17:30 TuCT1.4

Reconstruction of Impact on Composite Airfoil Segment Using Piezoelectric Sensors

Zemcik, Robert — Univ. of West Bohemia
Bartosek, Jan — Univ. of West Bohemia
Lasova, Zuzana — Univ. of West Bohemia
Kroupa, Tomas — Univ. of West Bohemia

The presented study focuses on structural health monitoring technique with the goal to predict induced damage and assess the residual strength of impacted structure. Herein, the primary problem of impact force reconstruction is investigated on an airfoil segment (part of a propeller blade for a large fan) made of laminated composite with attached piezoelectric transducers. The segment is made from carbon textile using RTM technology. The reconstruction is based on deconvolution method and transfer functions. The efficiency of the proposed approach is tested for all impact locations used to calibrate the system, and also on several additional locations. The interpolation of transfer functions is used to increase the accuracy of the reconstruction.

17:30-17:50 TuCT1.5

Cure Monitoring and SHM of Carbon Fiber Reinforced Polymer Part I : Electrical Impedance Measurement and Analysis

Mounkaila, Mahamadou — LAAS

The high-performance composite materials based on carbon fiber are increasingly used in critical security areas (aeronautics, civil engineering). Thus, it is important to know their characteristics during curing process or their use. We present herein a technique of electrical impedance analysis to extract some specific material properties (resistance and capacitance) in order to know its behavior. As the material microstructure contains a conductor part (carbon fiber) and an insulator part (resin), a model of electrical conduction in the material was established by using a network of a resistance RP connected in parallel with a capacitance CP (impedance Z). Spectroscopic impedance analysis of the studied samples informs about the intrinsic properties of material and shows a sensitivity of these electrical parameters (RP, CP) according to the curing cycle. Then the sensitivity to some physical parameters (temperature, deformations, etc.) will be demonstrated in order to know the health of the material, thus ensure Structural Health Monitoring (SHM).

17:50-18:10 TuCT1.6

Cure Monitoring and SHM of Carbon Fiber Reinforced Polymer Part II : Multi-Physical Correlations

Sassi, Sonia — ICA-TOULOUSE

This article concerns the on-line monitoring of the manufacturing of composite materials using an ad-hoc instrumentation (based on electrical impedance measurements) considering the material it-self as a sensor. The material (made of T700 carbon /M21 epoxy prepegs) is here used as a sensor but could be also used as an actuator. Furthermore the instrumentation was developed to stay in the heart of the structure for SHM purpose. It is proposed to measure the changes in the electrical impedance of the composite material during curing. The changes of the resistance Rp and the capacitance Cp during curing are then correlated to rheological parameters linked to the progress of the curing reaction. The present paper exposes the estimation of electrical and rheological properties of the material using the measurements of Rp according to a specific curing cycle.

TuCT2 — GH Room

Guided Waves III (Regular Session)

Chair: Su, Zhongqing — The Honk Kong Pol. Univ.

Co-Chair: Treyssède, Fabien — IFSTTAR

16:10-16:30 TuCT2.1

Investigation on High-Order Harmonic Generation of Guided Waves Using Local Computation Approaches: Theory and Comparison with Analytical Modelling

Radecki, Rafal — Department of Mechanical Engineering, The Hong Kong Pol.
Leamy, Michael — G. W. Woodruff School of Mechanical Engineering, Georgia Inst.
Staszewski, Wieslaw J. — Department of Robotics and Mechatronics, AGH Univ. of Science and Technology, Kraków
Uhl, Tadeusz — AGH Univ. of Science and Technology, Kraków
Su, Zhongqing — The Hong Kong Pol. Univ.
Cheng, Li — Department of Mechanical Engineering, The Hong Kong Pol.
Packo, Pawel — Department of Robotics and Mechatronics, AGH Univ. of Science and Technology, Kraków

The paper presents numerical modelling approaches for computing wave propagation and high-order harmonic generation in nonlinear media. This approach is based on two local computational methods, i.e. the Local Interaction Simulation Approach and the Cellular Automata for Elastodynamics. Discretized numerical dispersion curves are calculated and used for the entire analysis. A quantitative study of velocity matching between the fundamental and high-order harmonic waves is performed. The influence of propagation distance on the magnitude of high-order harmonics is evaluated for various excitation frequencies. Numerical analyses show divergence from analytical models for dispersion curves, especially for high-order harmonics. These preliminary investigations demonstrate that the numerical modelling approaches applied can be used effectively for nonlinear wavefield analysis; however, numerical instead of analytical spectral characteristics should be considered.

16:30-16:50 TuCT2.2

A Sparsity Promoting Algorithm for Time of Flight Estimation in Guided Waves - Based SHM

De Marchi, Luca — Univ. of Bologna
Moll, Jochen — Goethe Univ. of Frankfurt
Marzani, Alessandro — Univ. of Bologna

Ultrasonic Guided Waves (GW) are employed by many Structural Health Monitoring (SHM) systems. In plate-like components, GW based defect detection and localization is typically achieved through multiple piezoelectric transducers arranged in different array configurations. In active methods, one or more actuators are used to generate GWs and the sensors work as wave detectors. Defects can be detected and localized from the wave scattering that they generate. To increase the precision of localization approaches, it is important to minimize the uncertainty in the estimation of the time of flight (ToF) of the waves scattered by the defect. Such task is complicated by the dispersive and multimodal nature of ultrasonic GW propagation. In this work, we analyse two algorithms to extract the ToF from waveforms acquired with a Scanning Laser Doppler Vibrometer (SLDV). The algorithms outputs are used to feed beamforming procedures to image cracks with various orientations.

16:50-17:10 TuCT2.3

Damage Detection, Localization and Size Estimation Using Broadband Correlation-Based Imaging

Ostiguy, Pierre Claude — GAUS – Dept Mechanical Engineering – Univ. of Sherbrooke Sh
Quaegebeur, Nicolas — GAUS – Dept Mechanical Engineering – Univ. of Sherbrooke Sh
Masson, Patrice — GAUS – Dept Mechanical Engineering – Univ. of Sherbrooke Sh

Imaging approaches based on guided waves aim at detecting, locating and estimating the damage severity on a structure. The actual limitations of imaging approaches are that (1) sensitivity do damage depends on frequency used, and (2) the damage growth can be observed but not yet quantified. This paper presents a robust methodology for damage imaging and size estimation using reconstructed broadband signals, where measurements of the transfer function between each emitter and receiver are made using a sub-band decomposition strategy. Pristine transfer function is subtracted from damage transfer function signature. Imaging is conducted using a correlation-based approach (Excitelet), and dispersion compensation with reconstructed broadband signals. The approach is validated experimentally on a 1.54 mm thick aluminium plate, where only three piezoceramic transducers are bonded on the structure. Measurements are taken for two artificial damage of 13 mm and accurate detection and dimensioning is achieved. The analysis of the transfer functions using the A0 mode shows that the wavelengths reflecting most of the energy are associated to the damage size within an accuracy of one millimetre.

17:10-17:30 TuCT2.4

Application of Multifrequency Spectral Method of Lamb Waves for Structural Health Monitoring of Composite Laminates

Pogorielov, Oleksandr — Independent Res.

Article is devoted to a new approach of the implementation of the method of Lamb waves applied for SHM. The new approach is to borrow the pulse synthesis method from the techniques of microwave frequency. The approach consists of the application of multi-frequency band measurements and the integral spectral transformations to interpret the results. The method is focused on the application for the reinforced composite materials to solve the SHM problems. The nonlocal eigenoscillation method and a local method of Lamb waves can be implemented within the framework of this approach and measuring equipment. The article describes the method and the experimental results that demonstrate the possibilities of the method.

17:30-17:50 TuCT2.5

Stationarity-Based Approach for Lag Length Selection in Cointegration Analysis of Lamb Wave Data

Dao, Phong B. — Department of Robotics and Mechatronics, AGH Univ. of Science and Technology, Kraków
Staszewski, Wieslaw J. — Department of Robotics and Mechatronics, AGH Univ. of Science and Technology, Kraków

It is well known that when cointegration is used for the analysis of data in structural health monitoring applications, the choice of lag length has strong influence on damage detection results. The paper demonstrates how this problem could be solved. The solution utilizes the inversely proportional relationship between damage severity and stationarity of cointegration residuals. The method is validated using Lamb wave data from a structure exposed to temperature variations. The experimental results show that the proper lag length selection is essential and this can be achieved with the appropriate statistical analysis.

17:50-18:10 TuCT2.6

Precision of Imaging Algorithms Based on ToF Estimation of Guided Waves

Le Duff, Alain — Groupe ESEO
Quaegebeur, Nicolas — GAUS - Univ. of Sherbrooke
Ostiguy, Pierre Claude — GAUS – Dept Mechanical Engineering – Univ. of Sherbrooke Sh
Masson, Patrice — GAUS – Dept Mechanical Engineering – Univ. of Sherbrooke Sh

This article addresses the problem of damage localization in an isotropic plate using guided waves. This localization is based on Time of Flight measurements thanks to an array of three sensors arbitrarily distributed on the structure. First, an exact solution of the Time of Flight problem is proposed. Then, the exact analytical Cramer-Rao Bounds, expressed as a function of the actuator and the sensors locations and the Signal to Noise Ratio of the measured signals, are presented. Statistical performance of the method is illustrated by means of Monte-Carlo simulations and compared to the Cramer-Rao Bounds. Time delays are estimated both with an Hilbert Transform method and a Maximum Likelihood Estimator. Results show that the performance of the algorithms is in agreement with the theoretical variance if Signal to Noise Ratio is low enough. These Cramer-Rao Bounds provide a way to determine the optimal position of a sensor array for a given a-priori accuracy.

TuCT3 Room I

SHM Applications I (Regular Session)

Chair: Bernal, Dionisio — Northeastern Univ.
Co-Chair: Brouns, Jordan — IFSTTAR

16:10-16:30 TuCT3.1

Structural Health Monitoring Solutions for Power Plants

Rudolph, Jürgen — AREVA GmbH
Bergholz, Steffen — AREVA GmbH
Jouan, Benoît — AREVA GmbH

Structural health monitoring solutions gain in importance not only as part of the ageing management of nuclear power plant components but also in the context of conventional power plants and renewables such as wind power plants. Consequently, lots of operators have to deal with demanding security requirements to ensure the safe operation of power plants and to cope with plant lifetime extension (PLEX) related issues. AREVA disposes of a long tradition in the development of structural health monitoring (SHM) solutions. Nuclear and conventional power plant applications require the qualified assessment of measured thermo-mechanical loads. The methodology is transferable to mechanical loading conditions such as those of wind energy plants. The core challenge is the identification and qualified processing of realistic load-time histories. In terms of the nuclear industry, the ageing management of power plant components is nowadays a main issue for all actors: states, regulatory agencies, operators, designers or suppliants. As regards fatigue assessment of nuclear components stringent safety standards imply the consideration of new parameters in the framework of the fatigue analysis process: new design fatigue curves, consideration of environmental fatigue (EAF) parameters and stratification effects. In this general context

AREVA developed the integral approach AREVA Fatigue Concept (AFC) with new tools and methods in order to live up to operators expectations: Simplified Fatigue Estimation (SFE), Fast Fatigue Evaluation (FFE) and Detailed Fatigue Check (DFC). Based on real measured thermal loads and superposed mechanical loads the Fast Fatigue Evaluation (FFE) process allows a highly automated and reliable data processing.

16:30-16:50 TuCT3.2

Manual UT vs Permanently Installed Sensors

Cegla, Frederic — Imperial Coll.

Ultrasonic NDE and inspection techniques are often strongly influenced by procedures and human factors as well as the underlying physics of the measurement technique. Automation of some of the acquisition tasks potentially eliminates some sources of error. The NDE group at Imperial College London has developed permanently installed ultrasonic thickness monitoring tools which allow frequent acquisition as well as automatic evaluation of ultrasonic wall thickness estimates. Several thousand sensors have been deployed in field and some of the outcomes from site data will be presented. The factors that influence the automated system's performance will be discussed in this paper with a view to giving an indication of the different aspects that are important when trying to evaluate thickness measurements as a function of time. The automated thickness measurements are then compared to measurements that are collected with manually collected measurements using the conventional inspection approach.

16:50-17:10 TuCT3.3

Monitoring of Vibrations for the Protection of Architectural Heritage

Del Grosso, Andrea Enrico — Univ. of Genoa
Basso, Paolo — D'Appolonia S.p.A.

A peculiar aspect in Structural Health Monitoring is concerning the monitoring of vibrations in urban environments that may affect the integrity and conservation of architectural heritage buildings. The paper reviews the main issues on the subject and takes into consideration the problem of measuring and interpreting vibrations in buildings. The use of numerical models is also discussed and a practical case study is finally presented.

17:10-17:30 TuCT3.4

Instantaneous Damage Identification and Localization through Sparse Laser Ultrasonic Scanning

Park, Byeongjin — KAIST
Sohn, Hoon — Professor, Department of Civil and Environmental Engineering, KA

This study proposes an instantaneous damage identification and localization technique through sparse laser ultrasonic scanning. First, an inspection path is selected from an intended inspection region of a target structure. Then, ultrasonic waves are generated at one endpoint A of the path with a pulse laser and the corresponding ultrasonic responses are obtained at the other endpoint B with a laser Doppler vibrometer (LDV). Second, ultrasonic waves generated at point B are measured at point A reciprocally. Once two reciprocal ultrasonic signals are obtained, a damage index (DI) representing the violation of the linear reciprocity on the path is computed by comparing the reciprocal signals. DI will increase when the path passes through damage as this generates nonlinearity to the path. Third, the aforementioned steps are repeated for all predetermined inspection paths within the inspection region by scanning both pulse laser and LDV. Finally, the paths associated with high DI values are identified as damage regions. The effectiveness of the proposed sparse scanning technique is validated using an aluminium plate with a fatigue crack and a composite plate with an impact induced delamination.

17:30-17:50 TuCT3.5

A Methodology of Analysis for a Critique Interpretation of the Data Acquired from Monitoring Systems of Historical Buildings

Trombetti, Tomaso — Univ. of Bologna
Baraccani, Simonetta — Univ. of Bologna
Palermo, Michele — Univ. of Bologna
Silvestri, Stefano — Univ. of Bologna
Gasparini, Giada — Univ. of Bologna
Dib, Antoine — Univ. of Bologna

Preserving historical buildings is essential in the safeguard of the cultural heritage of any country. The need to carry out structural analysis with non-destructive methods gave structural monitoring a widespread fame in the world of diagnosis and control of historical buildings. The aim of this study is to introduce a standardized approach for the analysis of the data acquired from a monitoring system of an historical building, through the definition of specific reference quantities (extrapolated from the recorded time series) able to characterize the main features of the structural response and the preliminary identification of the order of magnitudes of these quantities. It is assumed that the recorded time series may be decomposed into two fundamental components: the first one related with the natural actions and characterized, in absence of extreme events, by a substantial periodic behavior, the second one related to the other factors such as the evolution of the state of the structure due to material degradation, soil settlements and others. Exploiting the properties of periodic functions, one may identify these reference quantities, which are based on the year and the day variability and allow to monitor the evolution of the phenomena under observation. These reference quantities may be collected in a database and may become fundamental for comparing the response of similar buildings. This methodology has been applied to the data obtained from the monitoring system of the Cathedral of Modena

17:50-18:10 TuCT3.6

Damage Detection Strategies in Structural Health Monitoring of Overhead Power Transmission System

Skarbek, Lukasz — Faculty of Electrical and Control Engineering, Gdańsk Univ.
Zak, Arkadiusz — Faculty of Electrical and Control Engineering, Gdańsk Univ.
Ambroziak, Dominik — Faculty of Electrical and Control Engineering, Gdańsk Univ.

Overhead power transmission lines, their supporting towers, insulators and other elements create a highly distributed system that is vulnerable to damage. Typical damage scenarios cover cracking of foundation, breakage of insulators, loosening of rivets, as well as cracking and breakage of lines. Such scenarios may result from various factors: groundings, lightning strikes, floods, earthquakes, aeolian vibrations, conductors galloping, icing, and also include sabotage or terrorist attacks. It is impossible to monitor every kilometer of power lines, especially when they pass through inaccessible areas like mountains or swamps. This implies that many of the damage scenarios mentioned remain unnoticed or are detected too late. In most cases damage of the power system leads to instantaneous blackouts of small regions, but huge blackouts are also possible. Every power outage causes a great financial loss, threats to hospital patients, public transport or even national security. The paper presents a brief introduction to the field of damage detection in transmission towers, emphasizing indexes calculated by simple processing and using artificial neural networks. Problem of line sag monitoring is also mentioned. The paper ends with a short report on possible financial impacts of an SHM system for electric energy transmission and distribution system.

TuCT4 — Room J
Image Processing (Regular Session)

Chair: Pakrashi, Vikram — Univ. Coll. Cork

Co-Chair: François, Marc — Faculté des Sciences de Nantes, Lab. GeM

16:10-16:30 TuCT4.1

A Comparison of Image Based 3D Recovery Methods for Underwater Inspections

O'Byrne, Michael — Trinity Coll. Dublin
Schoefs, Franck — Univ. of Nantes

Pakrashi, Vikram — Univ. Coll. Cork
Ghosh, Bidisha — Trinity Coll. Dublin

Offshore structures can be subjected to millions of variable amplitude load cycles during their service life which is the primary cause of structural deterioration. Such fatigue loading is exacerbated by marine growth colonization which changes the surface roughness characteristics and increases the diameter of structural members. Having an accurate knowledge of these parameters is essential for analyzing the increased hydrodynamic forces acting on the structure. This paper addresses the issue of acquiring shape information by comparing two popular classes of image based shape recovery techniques; stereo photography and Structure from Motion (SfM). Stereo photography utilises a dual camera set-up to simultaneously photograph an object of interest from slightly different viewpoints, whilst SfM methods generally involve a single camera moving in a static scene. In this paper, these techniques are performed on a controlled shape in an underwater setting, as well as synthetic data which allows for an irregular shape typical of marine growth to be tested whilst still having knowledge of the exact geometrical shape. The results reveal that the self-calibrated stereo approach fared well at getting an appropriately scaled full metric reconstruction, whilst the SfM approach was more susceptible to breaking down.

16:30-16:50 TuCT4.2

Local Strain and Damage Measurements on a Composite with Digital Image Correlation and Acoustic Emission

Flament, Camille — Ec. Centrale de Lyon
Salvia, Michelle — Ec. Centrale de Lyon
Berthel, Bruno — Ec. Centrale de Lyon
Crosland, Gerard — VALEO Matériaux de Friction

Certain material have a non linear stress-strain behaviour under monotonic loading. This non linearity can be caused by continuous damage. To highlight this phenomena simultaneous strain and damage measurements should be taken. When working with inhomogeneous or anisotropic materials, such as continuous fibre composites, local information is essential. On the one hand, full field strain data can be obtained with Digital Image Correlation (DIC). On the other hand Acoustic Emission (AE) recording is a method that can be used for damage monitoring and location if, at least two sensors are placed on the specimen under loading. The aim of this work is to combine these two techniques to correlate strain measurements and damage location on a complex composite material during a monotonic tensile test. This composite is a continuous fibre reinforced friction material used in car clutches. These measures were used to understand the non linearity of the stress-strain curve of the as received material as well as detect volume damage after thermal cycling.

16:50-17:10 TuCT4.3

Monitoring of Inflatable Structures by Using Virtual Image Correlation

François, Marc — Faculté des Sciences de Nantes, Lab. GeM
Bloch, Alexis — GeM, Faculté des Sciences de Nantes
Thomas, Jean-Christophe — GeM, Faculté des Sciences de Nantes
Flamand, Olivier — CSTB

The VIC method consists in finding the analytical contour by fitting the boundary of the objet in the picture as best as possible. It has been used for shape measurement of an inflatable beam submitted to a transverse wind in the CSTB wind tunnel in Nantes. The principle and the main equations of the method are recalled. The originality of this study is that the equation of the contour is deduced from the analytical expression of the deformation of an inflatable beam under a distributed constant (aerodynamic) load. This acts as a powerful mechanical-based filter and leads straightforwardly to the researched mechanical properties of the textile membrane.

17:10-17:30 TuCT4.4

Vehicle-Barrier Tracking of a Scaled Crash Test for Roadside Barrier Design

Amato, Giuseppina — Queen's Univ. Belfast
Ghosh, Bidisha — Trinity Coll. Dublin
Simms, Ciaran — Trinity Coll. Dublin

In this paper the tracking system used to perform a scaled vehicle-barrier crash test is reported. The scaled crash test was performed as part of a wider project aimed at designing a new safety barrier making use of natural building materials. The scaled crash test was designed and performed as a proof of concept of the new mass-based safety barriers and the study was composed of two parts: the scaling technique and of a series of performed scaled crash tests.

The scaling method was used for 1) setting the scaled test impact velocity so that energy dissipation and momentum transferring, from the car to the barrier, can be reproduced and 2) predicting the acceleration, velocity and displacement values occurring in the full-scale impact from the results obtained in a scaled test.

To achieve this goal the vehicle and barrier displacements were to be recorded together with the vehicle accelerations and angular velocities. These quantities were measured during the tests using acceleration sensors and a tracking system. The tracking system was composed of a high speed camera and a set of targets to measure the vehicle linear and angular velocities. A code was developed to extract the target velocities from the videos and the velocities obtained were then compared with those obtained integrating the accelerations provided by the sensors to check the reliability of the method.

17:30-17:50 TuCT4.5

Surfaces Reconstruction Via Inertial Sensors for Monitoring

Saguin Sprynski, Nathalie — CEA Leti
Jouanet, Laurent — CEA Leti
Lacolle, Bernard — LJK - UJF
Biard, Luc — LJK - UJF

This document deals with the new capabilities of monitoring via the surface reconstruction of stuctures with sensors' arrays systems. Indeed, we will detail here our new demonstrator composed of a smart textile equipped with inertial sensors and a set of processings allowing to reconstruct the shape of the textile moving along time. We show here how this new tool can provide very useful information from the structures.

17:50-18:10 TuCT4.6

Effects of Enhanced Image Quality in Infrastructure Monitoring through Micro Aerial Vehicle Stabilization

Kuo, Chung-Hsin — Saarland Univ.
Kanlanjan, Sébastien — Univ. of Valenciennes
Pagès, Louis — Univ. of Valenciennes
Menzel, Hanadi — Saarland Univ.
Power, Sascha — Saarland Univ.
Kuo, Chen-Ming — Saarland Univ.
Boller, Christian — Saarland Univ.
Grondel, Sébastien — Univ. of Valenciennes

Traditional monitoring of large infrastructure such as towers of churches or for cooling, chimneys or any other type of tall buildings can require extreme effort and hence become very risky and costly since it mainly requires people to be moved around. An interesting alternative in that regard is the use of rotary wing micro aerial vehicles (MAV) equipped with sensors such as digital cameras to capture series of images and stitch them on to a 3D model. However the images recorded always have "noise" which is caused by the vehicle movement during the inspection process. Experiments have been carried out using a multi-rotor copter with an off the shelf camera for building inspection (aerial photography) for years. To validate the effect of MAV flight stabilization a vector thrust principle recently developed at Saarland University (patent application in progress) has been implemented onto a MAV for flight stabilization. The paper describes the procedure of MAV-based infrastructure monitoring as well as the image stitching process in general before explaining the vector thrust design principle and the enhancements achieved in terms of image resolution and processing.

TuCT5	KL Room
Piezoelectric Sensing (Regular Session)	
Chair: Chang, Fu-Kuo	Stanford Univ.
Co-Chair: Cusano, Andrea	Univ. of Sannio

16:10-16:30 TuCT5.1

Towards Efficient Integrated Piezoelectric Transducers for Sensing, Actuation and SHM

Neumeister, Peter — Fraunhofer Inst. for Ceramic Tech. and Systems IKTS

Flössel, Markus — Von Ardenne Anlagentechnik GmbH

Eßlinger, Sophia — Fraunhofer Inst. for Ceramic Tech. and Systems IKTS

Gebhardt, Sylvia — Fraunhofer Inst. for Ceramic Tech. and Systems IKTS

Schönecker, Andreas — Fraunhofer Inst. for Ceramic Tech. and Systems IKTS

The present contribution investigates the piezoelectric performance of novel ceramic modules for SHM applications. The multilayer technique is used to embed piezoceramic plates into an encasement made of low temperature cofired ceramics, including electric wiring and electronics. With the help of finite element analyses, the piezoelectric properties are derived from static bending experiments using large signal electrical loads. A significantly increased piezoelectric coupling is found for the embedded piezoelectric elements compared to single elements. This finding is explained by the particular stress state in the embedded piezoelectric element due to the fabrication process.

16:30-16:50 TuCT5.2

SHM for RC Structure Using PZT Rebar Active Sensing System – Numerical Study

Wu, Fan — Shanghai Jiao Tong Univ.

An active sensing diagnostic system using PZT for SHM of reinforced concrete structure has been currently under the investigation. The system consists of steel reinforcement bar (rebar) and PZTs bonded on the rebar surface. Test results show that the system can detect damages of the structure. To fundamentally understand the damage algorithm and therefore to establish a robust diagnostic mechanism, an accurate Finite Element Analysis (FEA) for the system has been performed. Reinforced concrete structure with a PZT bonded rebar is also simulated and analyzed with commercial FEA software. Simulation analysis is further made for concrete structure with different debonding damage sizes, and the results are validated with test ones. A series of signal optimization have been done through simulated parametric studies. Improvement of PZT installation has been discussed and the relationship between improvement and signal output has been established. By increasing both sizes of PZT actuators and sensors, the output signals increase linearly with PZT size and by several times.

16:50-17:10 TuCT5.3

Temporal Information of Linear and Nonlinear Lamb Waves for Fatigue Damage Localization: Analysis and Synthesis

Hong, Ming — The Hong Kong Pol. Univ.

Su, Zhongqing — The Hong Kong Pol. Univ.

Lu, Ye — Monash Univ.

Cheng, Li — The Hong Kong Pol. Univ. Hung Hom, Kowloon, Hong Ko

The time of flight (TOF) features of Lamb waves have been widely applied to locate gross damage in plate structures, which may greatly facilitate the localization process when a sparse sensor network is adopted. Through a time-frequency analysis, this study extends the usage of TOF to nonlinear Lamb waves, taking advantage of their higher sensitivity to small-scale fatigue damage. The precision and practicality of using temporal signal features of linear and nonlinear Lamb waves for fatigue damage characterization are compared. Case studies are conducted on aluminum plates through numerical simulation and experiments, where miniature piezoelectric wafers are networked into a sparse array for Lamb wave excitation and acquisition. A probability-based imaging algorithm is then proposed to visualize damage. Finally, the comparison has motivated a synthesized scheme for locating fatigue damage in various conditions, enhancing the adaptivity of structural health monitoring for real-world applications.

17:10-17:30 TuCT5.4

Hybrid Analytical-Spectral Method for the Modeling of Piezoelectrically Induced Waves in Plates

Vivar Perez, Juan Miguel — German Aerospace Center

Conventional numerical methods like the Finite Element Method fail to be efficient to model the propagation of ultrasonic guided waves in real structures. The required fine mesh and the high number of degrees of freedom when solving the wave propagation problem in the time domain lead to a high computational effort and a need of big memory storage capacity. On the other hand, analytical and semi-analytical methods, which offer fast and accurate results, can be only applied to relatively special geometries. In this contribution the use of an hybrid of analytical and spectral methods to model the propagation of elastic guided waves in plates is studied. Guided wave propagation excited by bonded piezoelectric transducers is simulated with this approach. The piezoelectric transducers are modeled using spectral methods and the response of the plate is calculated using analytical methods. The mathematical modeling of the transducer-plate interface bonding conditions is presented. These conditions are subsequently discretized using an hybrid analytic-spectral formulation in the frequency domain. The numerical accuracy of the obtained results is verified by comparison with other numerical methods such as the finite element method.

17:30-17:50 TuCT5.5

Check of Piezoelectric Transducers Using the Electro-Mechanical Impedance

Buethe, Inka — Univ. of Siegen

Moix-Bonet, Maria — DLR, Inst. of Composite Structures and Adaptative Systems -

Wierach, Peter — German Aerospace Center (DLR), Inst. of Composite Structures

Fritzen, Claus-Peter — Uni Siegen

The use of piezoelectric wafer active sensors in structural health monitoring systems implies the necessity to check these components for their functionality, especially when the sensors are embedded in the structure. This paper introduces two novel approaches based on electro-mechanical impedance measurements and compares them with a traditional approach based on the slope of the susceptance spectrum. The change of the resistance spectrum as well as the change of the correlation of susceptance spectra is established as damage indicator. In an experimental setup the mechanical durability of a piezoelectric patch transducer, directly embedded in the structure, was tested. It has been shown that this application of load causes partial sensor breakage, which, for this particular type of PWAS, can hardly be detected with the help of the susceptance slope. The good feasibility especially of the proposed damage indicator, based on correlation within the susceptance spectrum, is demonstrated.

17:50-18:10 TuCT5.6

Use of Two-Dimensional Stress in the Modelling of Guided Wave Generation by Circular Piezoceramics

Quaegebeur, Nicolas — GAUS - Univ. of Sherbrooke

Ostiguy, Pierre Claude — GAUS – Dept Mechanical Engineering – Univ. of Sherbrooke Sh

Masson, Patrice — GAUS – Dept Mechanical Engineering – Univ. of

Sherbrooke Sh

Traditional transducer design methods in Structural Health Monitoring based on Guided Wave propagation rely on the use of the pin-force model, assuming that a piezoelectric actuator can be modelled as a constant shear stress applied at its edge, whatsoever the frequency generated. However, the assumptions of this model are only validated for infinitely thin piezoelectric elements, weak coupling between the host structure and the transducer, and when the wavelength of the generated guided wave is above the size of the transducer. In this paper, a three-dimensional analysis of guided wave generation by a circular piezoceramic is proposed, taking into account the complex shear and normal interfacial stress profile between the transducer and the host structure. The complex stress profile is extracted from a Finite Element Model and then used as a parameter in the analytical propagation formulation. The influence of piezoceramic actuator dimension, thickness and host structure properties is then assessed numerically.

TuCT6 Room 200

Signal Processing for Monitoring of Wind Turbines (Invited Session)

Chair: Mevel, Laurent — Inria

Co-Chair: Döhler, Michael — Inria

16:10-16:30 TuCT6.1

Modal Parameter Estimation for Operational Wind Turbines (I)

Di Lorenzo, Emilio — LMS International
Manzato, Simone — LMS International
Peeters, Bart — LMS International
Marulo, Francesco — Univ. of Naples "Federico II"

Wind turbines are time-varying systems excited by loads due to the wind and to the interaction between blades, tower and drivetrain. Since it is very difficult to measure the loads, the modal identification procedure needs to rely only on the output measurement data. Operational Modal Analysis (OMA) is well suited for the estimation of modal parameters in several cases. One of the main conditions needed for its application is the linear time-invariance of the system. It is the case of parked wind turbines, but the requirement is violated in the case of operating wind turbines. Therefore, OMA technique needs to be adapted in order to be applied to linear time-variant systems. Alternatively, time-variant systems should be converted to time-invariant ones before applying the classical OMA. Multi-Blade Coordinate transformation (MBC) allows having information on the dynamic interaction between the nonrotating components and the rotor. The time periodic system is converted into a time invariant one. Conventional OMA technique can then be applied to estimate the modal parameters. First of all a multibody model of a wind turbine is considered and some assessments on how to combine numerical and experimental techniques for Structural Health Monitoring (SHM) of operating wind turbines are investigated.

16:30-16:50 TuCT6.2

Continuous Fatigue Assessment of an Offshore Wind Turbine Using a Limited Number of Vibration Sensors (I)

Iliopoulos, Alexandros — Vrije Univ. Brussel
Devriendt, Christof — Vrije Univ. Brussel
Guillaume, Patrick — Vrije Univ. Brussel
Van Hemelrijck, Danny — Vrije Univ. Brussel

Offshore Wind turbines are exposed to continuous wind and wave excitation that leads to high periodic stresses and strains at critical locations. This makes the structures prone to structural failure due to possible crack initiations and propagations. The continuous monitoring of the Wind Turbine is of utmost importance in order to assess the remaining lifetime and accumulative fatigue damage of the structure. Health monitoring of wind turbines is usually performed by collecting real-time operating data on a limited number of accessible locations using traditional sensors such as accelerometers and strain-gauges. When dealing with Offshore Wind Turbine though, most of the fatigue sensitive spots are inaccessible for direct measurements, e.g. at the muddline 30 meters below the water level. Response estimation techniques can then be used to estimate the response at unmeasured locations from a limited set of response measurements and a Finite Element Model. This paper makes use of a modal decomposition and expansion algorithm that allows for successful response prediction. The algorithm is validated using data obtained from a monitoring campaign on an offshore Vestas V90 3 MW wind turbine on a monopile foundation.

16:50-17:10 TuCT6.3

Loss of Rotor Isotropy As a Blade Damage Indicator for Wind Turbine Structure Health Monitoring Systems (I)

Tcherniak, Dmitri — Bruel and Kjaer Sound and Vibration Measurement

Modal-based damage features utilizing asymmetry of the rotor whirling mode shapes arising from rotor anisotropy are examined by Floquet analysis, and output-only modal analysis is applied to simulated vibrations of a rotating rotor.

17:10-17:30 TuCT6.4

Data Normalization for Foundation SHM of an Offshore Wind Turbine : A Real-Life Case Study (I)

Weijtjens, Wout — Vrije Univ. Brussel
Verbelen, Tim — Vrije Univ. Brussel
De Sitter, Gert — Vrije Univ. Brussel
Devriendt, Christof — Vrije Univ. Brussel

In this contribution the first results in the development of a SHM approach for the foundations of an offshore wind turbine will be presented. Key problems are the operational and environmental variability of the resonance frequencies of the turbine. This paper suggests a (non-)linear regression model to perform data normalization. Real-life data obtained from an offshore turbine on a monopile is used to validate the used model and to demonstrate the performance of the presented approach.

17:30-17:50 TuCT6.5

A Modular SHM-Scheme for Engineering Structures under Changing Conditions: Application to an Offshore Wind Turbine (I)

Häckell, Moritz — Leibniz Univ. Hannover, Inst. of Structural Analysis
Rolfes, Raimund — Leibniz Univ. Hannover, Inst. of Structural Analysis

Many countries worldwide and in Europe still have the goal of a future cut of CO2 emission in common. A shift from fossil to renewable energy source is the logical consequence. (Offshore) wind turbines ((O)WTs) play an important role in the so called "green" energy sector. An increasing number of remote offshore plants and an ageing fleet of onshore structures raise the demand of structural health monitoring (SHM) in this field. Guidelines still lack firm establishments and SHM is supposed to help assuring a safe operation and a possible extension of the lifetime.newline The work presented displays a modular SHM scheme applicable for engineering structures under varying environmental and operational conditions (EOCs). The procedure is applied to a 5MW OWT in the German bight, located in the test field textit{alpha ventus}. The integration into and application of the complete SHM scheme is presented through different condition parameters (CPs), machine learning (data classification) and hypothesis testing.

17:50-18:10 TuCT6.6

Robust Fault Detection Based on Multiple Functional Series TAR Models for Structures with Time-Dependent Dynamics (I)

Avendano-Valencia, David — Uiversity of Patras
Fassois, Spilios D. — Univ. of Patras

Vibration-based Structural Health Monitoring of operating wind turbines is challenging, as those structures are characterized by complex non-stationary response and are subject to environmental and operational uncertainties. FS-TARMA parameter based methods are ideal for this problem since they are capable of summarizing the non-stationary dynamics within a small parameter set. In this work, robust FS-TARMA parameter based fault detection methods are pursued by including several FS-TARMA models in the estimation of the statistical model used for posterior decision making. Different combination rules for the different FS-TARMA models are defined, analyzed and compared within the problem of vibration based fault detection on operating wind turbines using simulated data obtained from the FAST aeroelastic simulation code. Results demonstrate the improvement in terms of accuracy and reliability provided by the multiple model approach.

TuPoster Exhibition

Poster Session + Cocktail Party in Exhibition Space (Regular Session)

Chair: Le Cam, Vincent — IFSTTAR

Co-Chair: Mevel, Laurent — Inria

18:10-20:00 TuPoster

Nonlinear Ultrasonic Damage Detection for Fatigue Crack Using Subharmonic Component

Wang, Zhi — Department of Engineering Mechanics, Wuhan Univ. China
Qu, Wenzhong — Wuhan Univ.
Xiao, Li — Wuhan Univ.

Most modern structures operate under cyclic loading conditions, which may incubate fatigue cracks. In recent years, researchers have focused on more efficient procedures in SHM and damage detection using nonlinear vibration and nonlinear acoustic methods. Because superharmonics are also generated by the piezoelectric transducers and by the electronic equipment, these harmonics increase the background noise deteriorating the S/N ratio. The objective of this paper is to demonstrate the application of nonlinear ultrasonic subharmonic method for detecting fatigue cracks. A fatigue crack is qualitatively modeled as a single-degree-of-freedom (SDOF) system with non-classical hysteretic nonlinear interface force at both sides of the crack interface. The threshold of subharmonic is discussed and the influence of model parameters on the subharmonic resonance parameter condition is investigated. An aluminum beam with a fatigue crack is used to quantitatively verify the excitation voltage amplitude and frequency subharmonic resonance range. Two surface-bonded piezoelectric transducers are used to generate and receive ultrasonic wave signals. The experimental results demonstrate that the subharmonic components increase in magnitude with increasing amplitude of the input signal and can be used to accurately detect the fatigue cracks.

18:10-20:00 TuPoster

Decomposed Components of the Effective Temperature History and Their Correlation with the Variation of Stay Cable Force

Chen, Chien-Chou — Department of Construction Engineering, National Yunlin Univ.
Wu, Wen-Hwa — Nation Univ. of Science and Tech.
Liu, Chin-Yan — National Yunlin Univ. of Science and Tech.

Numerous studies have clearly demonstrated that the monitoring of cable force can be regarded as a royal road to diagnose the possible damages of cable-stayed bridges. The feasibility of this approach, however, deeply depends on accurate identification and filtration of the environmental temperature effect because it also significantly alters the cable force. A recent study by the authors discovered that an effective temperature defined by combining all the contributions from stay cable, bridge girder, and pylon is most strongly correlated with the cable force. Based on the above study, this work further applies the ensemble empirical mode decomposition to process the time histories of cable force and effective temperature for an extradosed bridge located in central Taiwan. It is evidently observed that the cable force and effective temperature can both be categorized into daily variation, season variation, and high-frequency noise with the order of decreasing weight. Moreover, a time shift around 1 to 2 hours generally exists between the daily variation of both investigated quantities. The correlation analysis conducted for the data of all four seasons indicates that daily and season variations have to be distinguished for accurately evaluating the temperature effect on the variation of cable force.

18:10-20:00 TuPoster

Classifiation of Systems' Health Condition Using the New Adaptive Fuzzy-Based Feature Classification Approach AFFCA in Comparison to a Macro-Data-Based Approach

Schiffer, Sandra — Chair of Dynamics and Control, Univ. of Duisburg-Essen
Rothe, Sandra — Univ. of Duisburg-Essen
Baccar, Dorra — Univ. of Duisburg-Essen
Söffker, Dirk — Univ. of Duisburg-Essen

In this contribution a recently developed new modeling and classification approach to be used with streamed measurement data of industrial processes is applied. This briefly repeated approach can be used for condition-based maintenance or structural health monitoring. The approach is based on a fuzzy-like modeling using statistical features from training data. Based on the trained model classification can be realized allowing to distinguish unknown data sets to the given number of data classes each related to states. Beside the detailed illustration of the approaches to be used, the results applying the automated classification using the AFFCA approach are shown. As data, complex and problem-specific Acoustic Emission (AE) signals and also signals taken from the operating level (called hydraulic pressure data) are used and compared. The results show that based on the AE-based and also on the hydraulic pressure based AFFCA system's health state classification, the changes of the system can be detected very early and detailed.

18:10-20:00 TuPoster

Algorithms Hardware Implementation for Ultrasonic Data Processing in SHM System

Alonso-Alegre, Javier — Univ. of the Basque Country
Aranguren, Gerardo — UPV/EHU
Monje, Pedro M. — Electronic Design Group, Univ. of the Basque Country
Alcaide Pardo, Ángel — AERNNOVA Engineering Solutions Iberica S.A.
Ruiz, Mariano — Univ. Pol. de Madrid
Barrera, Eduardo — Tech. Univ. of Madrid
Garcia-corcuera, Adolfo — UPV/EHU

Nowadays, devices that monitor the health of structures consume a lot of power and need a lot of time to acquire, process, and send the information about the structure to the main processing unit. To decrease this time, fast electronic devices are starting to be used to accelerate this processing. In this paper some hardware algorithms implemented in an electronic logic programming device are described. The goal of this implementation is accelerate the process and diminish the information that has to be send. By reaching this goal, the time the processor needs for treating all the information is reduced and so the power consumption is reduced too.

18:10-20:00 TuPoster

Development of an Ultrasonic NDT System for Automated In-Situ Inspection of Wind Turbine Blades

Ye, Guoliang — Brunel Univ.
Neal, Ben — TWI Ltd
Boot, Alex — Brunel Univ.
Kappatos, Vassilios — Brunel Univ.
Selcuk, Cem — Brunel Univ.
Gan, Tat-Hean — Brunel Univ.

It is crucial to maintain wind turbine blades regularly, due to the high stress leading to defects or even damage. Conventional methods require shipping the blades to a workshop for inspection off-site, which is extremely time-consuming and very costly. This work investigates using pulse-echo ultrasound to detect internal damages in wind turbine blades without the necessity to ship the blades off-site. A prototype of 2D ultrasonic non-destructive testing (NDT) system has been developed and optimised for in-situ wind turbine blade inspection. The system is designed to be light weight so it can be easily carried by an inspector climbing onto the wind turbine blade for in-situ inspection. It can be operated in 1D A-Scan, 2D C-Scan or 3D Volume Scan. A software system has been developed to control the automated scanning and show the damage areas in a 2D/3D map with different colours so that the inspector can easily identify the damage areas. Experiments on Glass Fibre Reinforced Plastics (GFRP) and wind turbine blades (made of GFRP) samples showed that internal defects can be detected. The main advantages of this system are fully automated 2D spatial scanning and flaw displaying. It is potentially to be used for in-situ inspection to save maintenance time and hence considered to be economically beneficial for the wind energy industry.

18:10-20:00 TuPoster

Structural Health Monitoring of the Support Structure of Wind Turbine Using Wireless Sensing System

Lu, Kung-Chun — National Center for Res. on Earthquake Engineering
Peng, Heng-Chu — National Cheng Kung Univ.
Kuo, Yu-Shu — National Cheng Kung Univ.

The wind turbine heavily depends on the success of the support structure to resist the complicated environmental loading, especially for the offshore wind turbine. How to manage these wind turbines and monitor the structural safety becomes an urgent and important issue today. To monitor the support structural safety of wind turbine, a good understanding of the structural dynamic behaviors of wind turbine is preliminary requirement. Due to the complicate dynamic behaviors of realistic wind turbine which includes the effects of the soil-structure interaction (SSI), the machine operation loading, the wind interaction, etc..., both the numerical model study and the system identification are required to clarify these complicated structural behaviors. In this study, a wireless sensing system, NTU-WSU, is installed in an on-shore large wind turbine (GE 1.5MW) to collect the structural responses, and the system performance of the proposed NTU-WSU was also evaluated in this study. Three measurement layouts and two operational scenarios (operating and stop states of the wind turbine) are considered in this study to clarify the complicated structural dynamic behaviors. Two system identification approaches are adopted in this study to extract and verify the structural dynamic features, direct Fourier spectrum observation method and frequency domain decomposition (FDD) method.

18:10-20:00 TuPoster

Fundamental Study for Improvement of Estimation Accuracy Chloride Content Using Electromagnetic Waves Method

Uchida, Mami — Hosei Univ.
Nojima, Junichiro — Hosei Univ.
Mizobuchi, Toshiaki — Hosei Univ.

Chloride contents estimation using electromagnetic waves as a non-destructive method is effective to conceive degradation of concrete structure, no damage to actual structure and wide range condition would be easily evaluated. From the past study confirmed that the more chloride content is increased in the concrete, the less amplitude value of the electromagnetic waves get. Utilization of this characteristic enables to estimate chloride contents. Giving light on the attenuation characteristics of electromagnetic waves by material change of properties in concrete, and lead the fundamental equation which contributes to estimate chloride contents in concrete using electromagnetic waves. The estimation of chloride contents around reinforcing bars is reported

18:10-20:00 TuPoster

Comments on Structural Condition Analysis Based on Strain Measurements on Tripod Model

Opoka, Szymon — Inst. of Fluid-Flow Machinery PASci
Mieloszyk, Magdalena — Inst. of Fluid-Flow Machinery PASci
Derra, Szymon — Inst. of Fluid-Flow Machinery PASci
Ostachowicz, Wiesław — Pol. Acad. of Sciences, Inst. of Fluid Flow Machinery

Selected strain measurements performed on tripod model are described in this paper. Two problems are investigated here. First, the influence of bonding procedure and surface's curvature on initial strain in optical sensor is shown. It seems that initial strains due to bonding procedure and curvature can be respectively high in comparison with allowed strain range of the optical sensors. Second, the ability of strain sensors to indicate changes in boundary condition in one leg of the tripod is also shown. For tripod model changes in boundary conditions can be detected by sensors not necessarily located at the bottom part of the structure. Maximal shear strain can be valuable indicator of such changes.

18:10-20:00 TuPoster

Empirical Evaluation of the Impact of Data Pre-Processing on the Performance of Predictive SHM of Jet Engines

Loyer, Jean-Loup — Inst. Superior Tecnico - Univ. of Lisboa

We evaluate the impact of data pre-processing on the performance of predictive Structural Health Monitoring algorithm on a real case study involving dozens of jet engines. A simple robust four-step framework is designed to this effect, made of 1) outliers removal, 2) range scaling, 3) variable selection (either by "manually" evaluating variable correlations or by quantification of variable importance via random forests) and 4) evaluation of the predictive performance of a unique selected binary classifier (random forests). The results contrast with the intuition and the literature, since pre-processing raw data decreases predictive performance in half of the cases analyzed. The isolated influence of each of the pre-processing techniques rank in this order: important variables chosen through random forests has the highest positive impact, followed closely by variable scaling and outlier removal to a lower extent, while the "manual" variable selection via the correlation matrix exerts a slightly negative impact on predictive performance. The influence of combining pre-processing techniques is in line with the isolated influence of each technique. However, a detailed evaluation should be done for every application since these results might be due to the high data quality of aerospace engines or to the characteristics of random forests.

18:10-20:00 TuPoster

Multipeak Treatment of FBG Sensor Signal

Fernández Valderas, Rosario — CATEC - Center for Advanced Aerospace Tech.
Gutiérrez Vázquez, Nicolás — CATEC - Center for Advanced Aerospace Tech.
Lasagni, Fernando Adrián — CATEC - Center for Advanced Aerospace Tech.

Fiber Bragg Grating (FBG) sensors are increasing their importance in Structural Health Monitoring (SHM) of composite materials. They excel at reduced size, electromagnetic immunity and corrosion resistance. Furthermore, they are sensitive to strain and temperature, allowing the record of both parameters simultaneously. For the case of composite materials, FBG sensors can be bonded to part surface or embedded within the CFRP plies prior to curing. During the last one, embedded sensors are exposed to high pressure, vacuum and temperature. These conditions modify fiber optic section, causing the degradation of the reflected signal. The received spectrum widens and shows multiple peaks for the same sensor, making signal treatment a complex task. In this paper, an algorithm for peak recognition and monitoring is presented and validated. Embedded sensors included in different set-ups are study through different loading conditions during mechanical testing.

18:10-20:00 TuPoster

Fiber-Optic-Based Pressure and Residual Strain Monitoring in CFRP Bonding Process: Toward Realization of Secondary Bonding in CFRP Aircraft Structures

Minakuchi, Shu — The Univ. of Tokyo

Uhira, Keigo — The Univ. of Tokyo
Terada, Yutaka — The Univ. of Tokyo
Takeda, Nobuo — The Univ. of Tokyo
Saito, Nozomi — Mitsubishi Heavy Industries, LTD.
Takayuki, Shimizu — Mitsubishi Heavy Industries Ltd

Applying secondary bonding to places where carbon fiber reinforced plastic (CFRP) structures are conventionally bolted leads to significant cost and weight saving. However, it has yet to be utilized due to difficulties in quality control. Lack of local pressure in bonding process has detrimental influence on the quality of the cured adhesive layer. Pressure detection is hence a key for controlling bonding quality, and a new quality assurance technique is urgently needed. This study established a fiber-optic-based pressure detection and quality assurance technique for secondary bonding process. Fiber Bragg grating (FBG) optical sensors were embedded in the adhesive layer and the reflected spectrum changes were monitored during the pressurizing stage to detect the applied pressure. Thermal residual strain after curing was also evaluated for further quality assurance. Good correlation was observed between the cure pressure, the CFRP quality and the FBG sensor response, confirming the validity of the proposed technique.

18:10-20:00 TuPoster

Development of a Bio-Inspired Structural Health Monitoring System Based on Multi-Scale Sample Entropy

Tzu-Kang, Lin — National Chiao Tung Univ.
Liang Jui-chang, Sezoot00 — nctu civil engineering

A bio-inspired structural health monitoring (SHM) system based on multi-scale Sample Entropy (SampEn) is proposed in this paper. Recently, studies on entropy have shown that the healthy state of human can be evaluated by analyzing the measured electrocardiogram. As similar circumstance is also faced in the field of structural health monitoring, where the vibration signal of the structure can be measured by deployed sensors, a multi-disciplinary research is inspired and conducted. Numerical simulation has verified that the damage location and condition can be detected by the proposed SHM algorithm by processing the ambient vibration signal.

18:10-20:00 TuPoster

Waste Hydrogen Pipelines Monitoring in Modern Power Plant

Gasior, Pawel — Wroclaw Univ. of Tech.

Many production processes in chemical industry generates hydrogen as a by-product. Part of this is reused in the other production process, however significant amount that is wasted can be used for energy production in stationary application. The goal of the project was development of the technology for safe exploitation of by-product hydrogen in chosen chemical plant. During the pilot project a small power plant with electrical capacity of 1 MW was designed and constructed. Two energetic technologies were applied. First use hydrogen and hydrocarbons as a fuel for combustion generators (leading technology), and the second one use purified hydrogen for fuel cells (complementary technology). A big attention was put on safety aspects and structural health monitoring of the most critical elements. An innovative technology based on distributed optical fiber based sensors for hydrogen pipeline integrity and leakage detection was applied. A temperature distribution along each of four supply gas pipelines (one with pure hydrogen to the fuel cell and three others with waste hydrogen to power generators) is permanently monitored.

Wednesday July 9, 2014

Wednesday July 9, 2014

Keynote Lectures (SHM & PHM) — Auditorium

Chair: Léger, Jean-Baptiste — Predict

08:30-09:00 — WeKN1L

Enabling Condition Based Maintenance for *Helicopters*

Hoffmann, Falk — Airbus Helicopters

Health and Usage Monitoring Systems (HUMS) for rotorcraft have been available since the early 1990s. The potential benefits of installing HUMS and transitioning towards Condition Based Maintenance (CBM), such as enhanced operating safety, increased availability, and reduced Direct Operating Costs (DOC) have been widely discussed, and their realization has been the subject of many research projects and operational trials in the past years. While HUMS installation and utilization has recently surged in the military environment, commercial applications have progressed at a much slower rate. Presently, HUMS are installed almost exclusively by operators in the Oil & Gas industry, to meet regulatory requirements that practically apply specifically to this industry. However, today there is a rising request coming also from other civil helicopter operators to exploit the full potential of HUMS. Helicopter manufacturers, like Airbus Helicopters, now have the challenging task to answer these demands with suitable HUMS and novel maintenance strategies that take advantage of condition and usage monitoring technologies.

To ensure the success of HUMS services, the specific needs of the end users must thoroughly be taken into account, which is the main difficulty due to the diversity of customers and their expectations towards the usage of helicopters. One of the main challenges in promoting Condition Based Maintenance lies in providing a cost-benefit justification. As the extent of the benefits is very difficult to predict in advance, a more generic and qualitative assessment must be established.
The keynote lecture will review the challenges that are faced in the application of HUMS and services built around such systems and will provide some answers and perspectives to next generation systems.

09:00-09:30 — WeKN2L

An Attempt at Diagnosing the Present and Prognosticating *the Future*

Dersin, Pierre — Alstom Transport

An attempt will be made at taking a look at the current state of the art of PHM and venturing some guesses as to where the discipline is headed; some modest suggestions will be offered.

The following issues will be addressed, or at least evoked, with a goal to stimulate discussion.

- Sensors: how to capitalise on technological progress for PHM? What progress is required?
- Intelligent data processing. What are the advances and the challenges on fault detection, fault separation (diagnostics) and prognostics?
- The various approaches to prognostics: reliability-based, model-based, statistics based; in particular, what are the challenges and potential breakthrough with adaptive modelling? How can model-based and statistical approaches be made truly complementary? How can slow degradations be sensed effectively enough to permit prognosis of remaining useful life?
- Data architecture and MMI aspects.
- What is meant by validation in the context of PHM?
- What are the added challenges in moving from equipment PHM to fleet PHM?

09:30 -10:00 — WeKN3L

Monitoring of Wind Energy Plants – Challenges and State of the Art

Fritzen, Claus-Peter — Siegen University

Kraemer, Peter — Woelfel Company

The technological development of wind energy plants (WEP) has made great progress during the past decades leading to larger and more powerful plants on- and offshore. However unexpected failure of components followed by drop out of energy production will lead to economic losses. An early detection of damage allows the operator to better plan necessary maintenance actions. Thus in parallel to the development of the WEP itself the necessity to monitor the central components of the plants has been recognized. Therefore monitoring systems are considered to play an important role to provide higher reliability of the energy production. Recently, another aspect came into the focus: older plants reach their approved lifetime of e.g. 20 years and the question arises how to prolong the lifetime and how to provide further safe operation?

The first monitoring systems for WEPs were related to operational data like wind speed, pitch angle, etc. as well as to the drive train components. These were supervised by means of condition monitoring systems (CMS). The CM-methods and systems for WEP have reached industrial standards today. These are described in CMS norms and guide lines and they found a broad acceptance in the industry.

The rapid growth of WEP installations over the world, especially of expensive WEPs with more than 3MW, since the year 2000 until now, brings the need of Structural Health Monitoring (SHM) of WEPs into the discussion. The idea was firstly welcomed with enthusiasm and different approaches transferred from civil and aeronautical engineering were originally adapted to monitor WEPs. The application of these approaches with the necessary measurement equipment to real WEPs during the last years have demonstrated interesting possibilities, but also their limitations and open challenges in the real field. Today the acceptance of WEP-SHM-Systems in the industry seems to be constricted, also the development of norms and guidelines on this sector progresses slowly. The reason for this current moderate acceptance is based in the higher investment on one side and the limited long-term robustness of the sensor systems and to some extent an insufficient significance of the delivered information regarding damage location and severity on the other side. Moreover uncertainties and tolerances in the manufacturing and design and the deeper understanding of possible damages and their consequences additionally contributes to a lack of target-oriented developments in the field. However, the key for the development of well accepted SHM-systems in the industrial field seems to be founded in a long-term transparent and trustful cooperation between the WEP component manufacturer, wind park operator and SHM-system designer.

This keynote presentation will show a short history of WEP development and the importance of the wind energy to the green energy supply today. In the following the significance of SHM and CMS for WEPs and the influences and mechanisms on their development will be explained.

The main part of the presentation will deal with the state of the art in CMS and SHM of wind energy plants in the research field and in industrial application. This will be accompanied by a depiction of some recent developments, possibilities and limitations of different kinds of approaches and systems for damage identification including also systems for identification of fatigue sources. In this context an example for a successful industrial realization of an SHM–system for rotor blade monitoring will be briefly described.

WeAT1 — Auditorium

OFS-4-SHM (Optical Fiber Sensors for Structural Health Monitoring) (Invited Session)

Chair: Ferdinand, Pierre — CEA LIST

Co-Chair: Akhras, George — Centre for Smart Materials and Structures - Royal Military Coll. of Canada

10:30-10:50	WeAT1.1

The Evolution of Optical Fiber Sensors Technologies During the 35 Last Years and Their Applications *in Structure Health Monitoring (I)*

Ferdinand, Pierre	CEA LIST

Since late 70s, (quasi-)distributed OFS have been developed, 12 countries producing 85% of the global effort. Since mid-80s, OFS has caught attention in many sectors where SHM is a matter of concern (civil engineering, composites, oil & gas, renewable energies, safety...). Today, OFS is became a strategic domain, especially in Asia, and China invests a lot since 7-8 years. Top 12 countries involved in SHM are, almost the same than in OFS, having published 80% of the total. Statistics proves that OFS is now the second sensing technology for SHM, and 2/3 concern the FBG sensing.

10:50-11:10	WeAT1.2

OFDR Distributed Strain Measurements for SHM of Hydrostatic Stressed Structures: An Application to High Pressure Hydrogen Storage Type IV Composite Vessels – H2E Project (I)

Maurin, Laurent	CEA, LIST, Optical Measurement Lab. F-91191 Gif-sur-Yvette
Ferdinand, Pierre	CEA LIST
Nony, Fabien	CEA
Villalonga, Stéphane	CEA

This paper deals with SHM of Composite Overwrapped Pressure Vessels (COPV) for high pressure hydrogen storage, based on embedded optical fibres and Rayleigh OFDR. Fibre Bragg Gratings in situ strain measurements on flat samples are first presented to assess the mechanical state of such anisotropic structures, and define a preliminary SHM criterion. Comparative fatigue tests were also performed on NOL rings to evaluate optical fibres mechanical impact, but their analysis in terms of energy revealed the difficulty to provide reliable results if the first cycles energy is not accurately controlled. Therefore, with no detrimental effect of the optical fibre embedment within the composite architecture, several instrumented pressure vessels were manufactured and tested. Accordingly, a mechanical criterion based on OFDR differential strain profiles analysis was defined to bring out internal defects, demonstrating the ability to early detect and locate internal flaws without requirements to inflate vessels at high pressures. Additionally, their sensitivity vs. pressure, analysed during a first burst test, provided additional valuable data about structure integrity. Finally, we define four complementary criteria based on in situ strain measurements to control pressure vessels for damage assessment, or any hydrostatic stressed structure since the underlying principles rely on strains proportionality vs. uniformly applied stresses.

11:10-11:30	WeAT1.3

Enhanced Brillouin Distributed Strain and Temperature Sensing for Structural Health Monitoring *in Industrial Applications (I)*

Rochat, Etienne	Omnisens
Nikles, Marc	Omnisens

The quest for energy is pushing fibre based monitoring into new frontiers. Similarly, structural health monitoring is also demanding advanced monitoring solutions. The combination of Brillouin optical time domain analyzers with specifically designed sensing cables and proper analysis tools is shown to be a perfect match for these challenging applications.

11:30-11:50	WeAT1.4

Dual-Fibre Bragg Grating Sensor for Simultaneous Temperature and Strain Sensing of Composite *Materials Manufacturing (I)*

Marin, Emmanuel	Univ. Jean Monet, LHC, Saint Etienne
Ouerdane, Youcef	Univ. Jean Monet, LHC, Saint Etienne

The Optical Fibre Sensors (OFS)-based monitoring of a composite part during its manufacturing process is presented. The sensor is made of a dual grating, i.e. a standard FBG and a Long Period Grating (LPG) photo-written at the same location, which provides accurate temperature and strain measurements, as well as a good spatial resolution. By using the differences of sensitivities of each grating, this dual-sensor is able to discriminate and to determine accurately temperature and strain. The dual-sensor has been tested during the curing of a pure epoxy resin, and a composite part obtained by Liquid Resin Infusion (LRI) process. Such dual-grating sensor is also very promising for SHM of composite parts.

11:50-12:10	WeAT1.5

Review on Optical Fiber Sensing Technologies for Industrical Applications at the NEL-FOST (I)

Yang, Minghong	Wuhan Univ. of Tech.
Li, Sheng	Wuhan Univ. of Tech.
Jiang, Desheng	Wuhan Univ. of Tech.

The research on engineering experiment is a key step in translating technical development to industrial application. According to our practical experience for more than 30 years and some applications of the fire alarm system, bridge, coal and power safety ensuring system, this paper reviews on engineering technique problems in the application of fiber optic sensor and their solutions, which may provide some references for wider industrial applications.

12:10-12:30	WeAT1.6

Microstructured Optical Fiber Bragg Grating Sensors for Structural Health Monitoring Applications (I)

Berghmans, Francis	Vrije Univ. Brussel
Geernaert, Thomas	Vrije Univ. Brussel - Brussels Photonics Team
Sonnenfeld, Camille	Vrije Univ. Brussel - Brussels Photonics Team
Sulejmani, Sanne	Vrije Univ. Brussel - B-PHOT Team
Luyckx, Geert	Univ. Gent - Department of Material Science and Engineeri
Lammens, Nicolas	Univ. Gent
Degrieck, Joris	Univ. Gent
Chah, Karima	Univ. of Mons - Electromagnetism and Telecom Department
Thienpont, Hugo	Vrije Univ. Brussel - Brussels Photonics Team

We first shortly review the state-of-the-art of microstructured optical fiber Bragg grating (MOFBG) sensors for structural health monitoring applications. We then focus on a specific microstructured optical fiber (MOF) design to which we refer as 'Butterfly' MOF. This fiber is highly birefringent and encodes the transverse strain into the spectral distance between the two Bragg peaks reflected by a fiber Bragg grating fabricated in this fiber. Since the birefringence of that MOF is not sensitive to temperature changes, the transverse strain measurement is independent of temperature variations. We subsequently discuss the potential of our MOFBG sensors for structural health monitoring related applications, including three-dimensional strain measurements within composite materials, cure monitoring and residual strain quantification following composite material manufacturing, shear stress measurements in lap joints and temperature insensitive vibration monitoring.

WeAT2	GH Room
Probabilistic SHM 1 (Invited Session)	
Chair: Zonta, Daniele	Univ. of Trento
Co-Chair: Schoefs, Franck	Univ. of Nantes

10:30-10:50	WeAT2.1

Reliability-Based Fatigue *Monitoring of Fracture Critical Structures*

Hernandez, Eric	Univ. of Vermont

This paper presents a finite element model-based state observer, which allows for accurate real time monitoring of stress, strain and their resultants at any location within an instrumented structure. The input to the proposed observer is: a finite element model of the structure, the spectral density matrix of the random field representing the loading environment and noise contaminated measurements of structural response. The resulting observer operates as a modified version of the original FEM of the structure driven by the measured response of the system. Using the rain flow cycle counting algorithm along with existing fatigue damage models, observer estimates are used to compute the number and amplitude of stress and strain cycles and the cumulative fatigue damage at critical points of the structure. These estimates can be used to obtain reliability estimates of the current state of the structure and prognosis under projected conditions.

10:50-11:10 WeAT2.2

Output-Only Statistical Time Series Methods for Structural Health Monitoring: A Comparative Study (I)

Vamvoudakis-Stefanou, Kyriakos J. — Univ. of Patras
Sakellariou, John — Univ. of Patras
Fassois, Spilios D. — Univ. of Patras

A comparative assessment of six well known Output-Only Statistical Time Series Methods (OO-STSMs) for Structural Health Monitoring (SHM) is presented via damage detection and identification in a GARTEUR type aircraft skeleton structure. A concise overview of the methods highlighting their principles is presented and their effectiveness for damage detection and identification is assessed via numerous experiments and various damage cases on the skeleton structure. What is more, issues such as the methods effectiveness based on local or remote vibration sensors as well as their computational complexity and ease of use are also investigated.

11:10-11:30 WeAT2.3

On Evaluating Monitoring Design Effectiveness (I)

Zonta, Daniele — Univ. of Trento
Cappello, Carlo — Univ. of Trento
Pozzi, Matteo — Carnegie Mellon Univ.
Glisic, Branko — Princeton Univ.

When designing a structure such as a bridge or a building, a civil engineer follows a well-established, rational procedure, whereby the performance of the design concept is predicted through structural analysis and quantitatively assessed with respect to the target performance. On the contrary, when an engineer designs a monitoring system, the approach is often heuristic with performance evaluation based on common sense or experience, rather than on quantitative analysis. In this paper, we describe a rational procedure for the design of monitoring systems, keeping in mind an analogy between structural and monitoring design. Whereas the structural design objective is to achieve stability with an appropriate level of safety, the object of monitoring is to acquire knowledge with an appropriate level of confidence. Herein, we illustrate the motoring design procedure with a simple practical example and discuss the possible metrics for evaluating the performance of a design concept, including: expected posterior accuracy of state parameters; relative entropy; probability of state mis-classification; value of monitoring information.

11:30-11:50 WeAT2.4

Probabilistic Structural Health Monitoring Using Passive-Only Damage Detection by Reciprocity of Green's Function Reconstructed from Diffuse Acoustic Fields (I)

Tippmann, Jeffery Dwayne — UCSD
Zhu, Xuan — Univ. of California, San Diego
Lanza di Scalea, Francesco — Univ. of California San Diego

A new passive damage detection approach is proposed using the reciprocity of the Green's function. This is achieved using a known method to reconstruct the forward and backward time-domain Green's functions between any two detection points in an acoustic diffuse field from the ensemble average of their cross-correlation functions. Damage is detected when the similarity between the forward and backward signals decreases due to the wave nonlinearities introduced by the discontinuity. Proof-of-principle results are discussed for an aluminum plate. A simulated diffuse field was created on the structures using a grid of independent random excitations. Aiming at damage detection, the observations were collected by two sets of experiments considering pristine structure and the case with simulated damages. Due to the lack of exact knowledge to the statistical distribution (most likely in a real world wind turbine blade), a bootstrapping parameter estimation was proposed in this study. To classify the cases with and without damages, a feature study was conducted based on multiple bootstrapping results. This damage detection method combining with the empirical statistical framework can be appropriately extended to any structure with significant acoustic or ultrasonic noise sources.

11:50-12:10 WeAT2.5

Examples of Damage Detection in Real-Life Settings Based on the Position of the Neutral *Axis (I)*

Sigurdardottir, Dorotea Hoeg — Princeton Univ.
Glisic, Branko — Princeton Univ.

The objective of the research discussed in this paper is to create damage detection methods based on universal parameters that are applicable to a large number and large variety of structures. Such methods, if successful, streamline the data analysis across large bridge networks and provide a basis for bridge managers to reach efficient, effective, yet safe and sustainable maintenance decisions. The first step towards this goal is to identify an appropriate parameter and validate that it can be used to detect damage in real-life settings. The centroid of stiffness is a universal parameter and its position in a cross-section can be evaluated for any load-carrying beam structure as the position of the neutral axis under conveniently chosen loads. Thus, a change in the position of the neutral axis within a cross-section can indicate change in the position of the centroid of stiffness, i.e., unusual structural behaviors. This research focuses on the neutral axis because of its universal applicability to beam-like structures and its direct correlation with unusual structural behaviors (e.g., damage). In this paper two examples of successful damage detection in real-life settings using the neutral axis are presented. In both cases the neutral axis is evaluated with long-gauge fiber-optic strain sensors installed on the structure in a parallel topology. The results show that the neutral axis is sensitive to damage and can be used to detect cracking and delamination in real-life settings, even years after the damage occurred.

WeAT3 Room I

Sensors III (Regular Session)

Chair: Cusano, Andrea — Univ. of Sannio
Co-Chair: Leduc, Dominique — GeM, Univ. of Nantes

10:30-10:50 WeAT3.1

The Real-Time Monitoring System for Metro Shield Tunnels: From Research to Application

Zhang, Jinfeng — Tongji Univ.
Liu, Wangsheng — Tongji Univ.
Zhao, Ming — Tongji Univ.

To the tunnel structure in soft soil foundation, damages are likely caused by large deformation, thus it is reasonable to measure deformation rather than to measure force or stress. Although there are many ways to obtain deformation information, measuring inclination of segments is one of the most effective ways, and it is convenient to derive the deformation from the inclination data. The method to obtain the variation of tunnel's diameter, the opening length of longitudinal connecting seams, the concrete stress of bolt and certain joint is studied in this paper by making use of the inclination variation of lining segments. This paper outlines a real-time monitoring system based on the internet of things which will be applied to an interval shield tunnel of Shanghai metro line 12 this year.

10:50-11:10 WeAT3.2

Compact Passive Wireless Corrosion Initiation Sensor That Can Be Installed in Existing Steel Reinforced Concrete

Perveen, Khalada	Univ. OF MANITOBA
Bhadra, Sharmistha	Univ. of Manitoba
Bridges, Greg	Univ. of Manitoba
Thomson, Douglas	Univ. of Manitoba

It is important to detect corrosion at the initiation stage for effective intervention to minimize the impact of corrosion. In this paper, we present a passive inductively coupled coil sensor that can detect corrosion potential with respect to a reference electrode. The sensor has a sensitivity of 0.13 kHz/mV. Several experiment results of different geometry of this sensor show that it can measure corrosion potentials with a resolution of less than 10 mV. The sensor is 25 mm in diameter and is intended to be installed by drilling a 25 mm or greater hole into existing concrete. The bottom of the hole is filled with mortar and the sensing electrodes are embedded within the mortar. Chlorides from the concrete diffuse into the mortar eventually reaching the electrodes leading to rise in the corrosion potential. This corrosion potential can be used indicate the level of chlorides within the concrete. The generated potential is a measure of the transfer of electron charge between the metal and its environment; here between the steel and the concrete. This is a property of the steel/concrete interface and not of the metal itself. Thus, it is impossible to determine the absolute value of this electrochemical potential and it is necessary to have a reference electrode to measure the potential difference of steel with respect to this reference electrode. The presented sensor can be fabricated at very low cost and is suitable for corrosion monitoring of existing structures.

11:10-11:30 WeAT3.3

Development of Wireless Ultrasonic Propagation Imaging System for In-Situ Aircraft Inspection

Chong, See Yenn	LANL-CBNU ENGINEERING Inst. KOREA, CHONBUK NATIONAL Univ.
Shrestha, Manish Man	DEPARTMENT OF MECHATRONICS ENGINEERING, CHONBUK NATIONAL Univ.
Lee, Jung-Ryul	DEPARTMENT OF MECHATRONICS ENGINEERING, DEPARTMENT OF AEROSPACE

Recently, laser ultrasonic propagation imaging (UPI) system has been greatly increased the attention of researchers for structural health management (SHM) application. The system has shown the promising results in the damage diagnosis of aircraft structures. However, the inspection was conducted in the experimental laboratory on a structure, which has been dissembled from an aircraft, which is not practical for the in-situ and large-scale aircraft structure inspection. This paper presents a wireless UPI system to realize the in-situ aircraft structure inspection. The system consists of a high-speed laser ultrasonic scanning system, a preamplifier-integrated piezoelectric (PZT) sensor network, an embedded wireless ultrasonic device (WUD), and a ground SHM system. In this work, four 8-channel analog multiplexers were used to connect the sensor nodes, which were installed along the webs of an aircraft, to the WUD with an integrated four-channel 12-bit analog-to-digital converter (ADC). The maximum sampling frequency of the ADC is 80 MHz. The WUD has an integrated Wi-Fi modem (802.11b/g/n). As further work, the UPI system will perform the inspection on a desired area with a selected local sensor node. Then, the acquired ultrasound will send to the ground SHM system to process for damage evaluation.

11:30-11:50 WeAT3.4

A Multifunctional Device for Multichannel EMI and Guided Wave Propagation Measurements *with PWAS*

Fritzen, Claus-Peter	Univ. of Siegen
Moll, Jochen	Goethe Univ. of Frankfurt
Chaaban, Rannam	Univ. of Siegen, Inst. of Mechanics and Control Enginee
Eckstein, Benjamin	EADS Innovation Works
Kraemer, Peter	Woelfel Beratende Ingenieure GmbH + Co. KG
Klinkov, Maksim	Achenbach Buschhütten GmbH
Dietrich, Gerhard	Univ. of Siegen, Inst. of Mechanics and Control Enginee
Yang, Cheng	Univ. of Siegen, Inst. of Mechanics and Control Enginee
Xing, Ke Jia	Univ. of Siegen, Inst. of Mechanics and Control Enginee
Buethe, Inka	Univ. of Siegen

This paper presents a multifunctional device enabling multichannel-measurements on piezoelectric elements and supports impedance and wave propagation measurements as well as the recording of additional information about environmental conditions within one experimental setup. This way of data acquisition enables the possibility of producing automated SHM systems according to the specific needs of the monitored structure including sensor assessment. A description of the hardware and software used in the process of data collection is presented and its applicability is demonstrated in some exemplary applications.

11:50-12:10 WeAT3.5

Crack Propagation Measurement Using a Battery-Free Slotted Patch Antenna *Sensor*

Yi, Xiaohua	Georgia Inst. of Tech.
Cho, Chunhee	Georgia Inst. of Tech.
Cook, Benjamin S.	Georgia Inst. of Tech.
Wang, Yang	Georgia Inst. of Tech.
Leon, Roberto T.	Virginia Pol. Inst. and State Univ.
Tentzeris, Manos M.	Georgia Inst. of Tech.

This research studies the performance of a battery-free wireless antenna sensor for measuring crack propagation. In our previous work, a battery-free folded patch antenna was designed for wireless strain and crack sensing. When experiencing deformation, the antenna shape changes, causing shift in electromagnetic resonance frequency of the antenna. The wireless interrogation system utilizes the principle of electromagnetic backscattering and adopts off-the-shelf 900MHz radiofrequency identification (RFID) technology. Following the same sensing mechanism, a slotted patch antenna sensor of smaller size is designed. The antenna detours surface current using slot patterns, so that the effective electrical length is kept similar as previous folded patch antenna. As a result, the sensor footprint is reduced and the antenna resonance frequency is maintained within 900MHz RFID band. To validate the sensor performance for crack sensing, a fatigue crack experiment is conducted on a steel compact-tension specimen. A slotted patch antenna sensor is installed at the center of the A36 steel specimen. For wireless interrogation, a Yagi reader antenna is placed 36 in. away from the antenna sensor to wirelessly measure the resonance frequency shift of the antenna sensor. The measurement is taken after every 10,000 loading cycles, till the specimen fails. Meanwhile, the length and width of the fatigue crack are also recorded. Finally, the resonance frequency shift of the antenna sensor is correlated with crack length and width at each loading stage.

12:10-12:30 WeAT3.6

Early-Age Concrete Strength Estimation Technique Using Embedded Piezoelectric Self-Sensing Impedance

Kim, Junkyeong	Sungkyunkwan Univ.
Kim, Ju-Won	Sungkyunkwan Univ.
Park, Seunghee	Sungkyunkwan Univ.

Recently, demands for the construction of Nuclear Power Plants (NPP) using high strength concrete (HSC) has been increased. However, HSC might be susceptible to brittle fracture if the curing process is inadequate. To prevent unexpected collapse during and after the construction of HSC structures, it is essential to confirm the strength development of HSC during the curing process. However, several traditional strength-measuring methods are not effective and practical. In this study, a novel method to estimate the strength development of HSC based on electromechanical impedance (EMI) measurements using an embedded piezoelectric sensor is proposed. The EMI of NPP concrete specimen was tracked to monitor the strength development. In addition, cross-correlation coefficient was applied in sequence to examine the trend of the impedance variations more quantitatively. The results confirmed that the proposed technique can be applied successfully monitoring of the strength development during the curing process of HSC structures.

WeAT4 Room J
Pattern Recognition I (Regular Session)

Chair: Samé, Allou — IFSTTAR
Co-Chair: Kullaa, Jyrki — Helsinki Metropolia Univ. of Applied Sciences

10:30-10:50 WeAT4.1

Wear-Oriented State-Of-Health Calculation and Classification Using Operating Data

Rothe, Sandra — Univ. of Duisburg-Essen
Söffker, Dirk — Univ. of Duisburg-Essen

Reliability and availability of technical complex and safety-critical systems are of increasing importance. The degree of wear as well as the quality of mechatronic systems are significant for the system reliability. To classify the machines state using easy-to-measure signals, two issues are important: filtering and interpretation of the data [1]. Core of this contribution is the development and first application of a simple, easy to use, easy to apply, and easy to handle algorithm to be used directly with industrial data or measurements from technical systems during operation. In this contribution a hydraulically driven machine part sliding over another is used as example. A connection between measured hydraulic data to the degree of wear of the lubricated surface is established to calculate information about the state of a sliding surface. As experimental data the time behavior of hydraulic pressure data is taken and filtered for better evaluation. To the further generation of suitably defined characteristics, the data are edited and analyzed. The results based on four measurements with two different operating conditions show that the developed approach allows a detailed judgment of wear-oriented state of health as part of a new structural health monitoring system.

10:50-11:10 WeAT4.2

Development of a Fault Detection Approach Based on SVM Apllied to Industrial Data

Rother, Astrid — ThyssenKrupp Steel Europe
Jelali, Mohieddine — Cologne Univ. of Applied Sciences
Söffker, Dirk — Univ. of Duisburg-Essen

In existing production plants, sensor systems and other sources provide information about the plant condition. This paper presents methods for how data can be conveniently summarized, treated, and evaluated to retain characteristic features and allocate them to certain faults respectively to use them for monitoring purposes. This work details the development of a method to be applied to selected data sets, and which then can be expanded for use in the real environment. This paper details a procedure developed for automated selection and processing to reduce the time exposure of qualified personnel. A number of possible methods of analysis were tested for their ability to point out conspicuous events, especially Wavelet Transformation for feature extraction and Support Vector Machines for classification. Data sets that are correlated to the different conditions of the system are used for training and testing. After training, the algorithm will be able to detect different faults in time. In today's practice, faults are analyzed after they have occurred. Applying the method, a major failure can be prevented by detecting contingency faults. Using real industrial data from the hot strip mill of ThyssenKrupp Steel Europe (TKSE), the developed approach will be tested offline for practical relevance.

11:10-11:30 WeAT4.3

Damage Estimation Using *Muli-Objective Genetic Algorihtms*

Shabbir, Faisal — Univ. of Engineering and Tech. Taxila, Pakistan
Omenzetter, Piotr — Univ. of Aberdeen

It is common to estimate structural damage severity by updating a structural model against experimental responses at different damage states. When experimental results from the healthy and damaged states are available, the updated finite element models corresponding to the two states are compared. Updating of these two models occurs sequentially and independently. However, experimental errors, updating procedure errors, modelling errors and parametric errors may propagate and become aggregated in the damaged model in this approach. In this research, a multi-objective genetic algorithm has been proposed to update both the healthy and damaged models simultaneously in an effort to improve the performance of the damage estimation procedure. Numerical simulations of a simply supported beam damaged at multiple locations with noisy mode shapes were considered and improved model updating results were confirmed. It was found that the proposed method is more efficient in accurately estimating damage severity, less sensitive to discretization as well as experimental errors, and gives the analyst an increased confidence in the model updating and damage estimation results.

11:30-11:50 WeAT4.4

Structural Health Monitoring*: A Quest towards the Use of Combined Approaches*

Buyukozturk, Oral — Massachusetts Inst. of Tech.
Long, James — Massachusetts Inst. of Tech.
Mohammadi Ghazi, Reza — Massachusetts Inst. of Tech.
Cha, Young-Jin — Massachusetts Inst. of Tech.
Chen, Justin — Massachusetts Inst. of Tech.
Smit, Dirk — Royal Dutch Shell

Over the past number of decades the structural health monitoring (SHM) research community has developed and published a large variety and number of methodologies for the purpose of detecting and locating damage in a structure using sensor measurement data. While almost all of these methods have demonstrated some degree of success in detecting damage, different approaches have differing costs, and corresponding tradeoffs in performance. Typical costs include computational effort, the development of an accurate structural model, or the collection of a large volume of data. Whether or not these costs are worth the investment depends on the specific SHM scenario. In this paper we analyze four different SHM methodologies, including model-based and data-based approaches, outlining their individual strengths and weaknesses, tradeoffs between cost and performance,and suggesting appropriate application areas for each. The efficacy of the methods is evaluated using data collected from a steel-frame laboratory structure.

11:50-12:10 WeAT4.5

Automatic Tuning of a Pipeline Faults Detection Algorithm

Villamizar, Rodolfo — Univ. Industrial de Santander
Camacho, Jhonatan — Univ. industrial de Santander
Perez, Oscar — Univ. industrial de Santander

This paper discusses the experimental results obtained by using a principal component analysis based algorithm in joint with Self Organizing Maps (SOM) Neural Networks for detection of damages in pipeline structures. Also, a differential evolutive algorithm is used for tunig the neural network parameters. A pipeline section test structure was instrumented with an active piezoelectric system in order to apply a high known frequency signal and to determine the base-line structural dynamical performance. Several piezoelectric sensors were located along the structure surface and damage features are obtained by processing the time vibrational dynamical response through principal component analysis. Q-statistic and Hotteling T2 indexes obtained from the PCA analysis are used to detect deviations of the current vibrational response respect to the undamaged one. Algorithm validation was achieved by using experimental data, obtained from a carbon steel tubing section, where damages were induced by adding masses to the structure. The obtained results indicate that is possible to identify and locate faults in pipeline structures and by using evolutive differential genetic technique improve the performance of the studied algorithm

12:10-12:30 WeAT4.6

Damage Detection in Metallic *Beams from Dynamic Strain Measurements under Different Load Cases by Using Automatic Clustering and Pattern Recognition Techniques*

Sierra, Julian	Univ. Pol. de Madrid
Torres Arredondo, Miguel Angel	Centre for Sensor Systems, Univ. of Siegen
Cabanes, Guenael	Paris XIII Univ.
Guemes, Alfredo	UPM
Mujica, Luis Eduardo	Tech. Univ. of Catalonia
Fritzen, Claus-Peter	Uni Siegen

In general, the change in the local strain field or global stiffness caused by damage in a structure is very small and the strain field tends to homogenize very quickly in the field close to the defect. Moreover, other environmental effects can fade the slight changes in the strain field. Only by comparing the response of the structure at several points some information about damage may be unveiled. By means of pattern recognition techniques based on the strain field, this task can be achieved. This is the basis of the strain measurements data-driven models. The main limitation of the strain field pattern recognition techniques lies in the susceptibility of the strain field to change depending on the load conditions. In the case of dynamic loads, this may reflect even a greater limitation. Robust automated techniques are required to manage these limitations. In first instance, automatic clustering techniques are needed so that data can be classified according to the load conditions and secondly, a dimensional reduction technique is needed in order to obtain patterns that often underlie from data. Within the context of this paper, a combination of Local Density-based Simultaneous Two-Level (DS2L-SOM) Clustering based on Self-Organizing Maps (SOM) and Principal Components Analysis (PCA) is proposed in order to firstly, classify load conditions and secondly, perform strain field pattern recognition. The clustering technique is the basis for an Optimal Baseline Selection. An experimental validation of the technique is discussed in this paper, comparing damages of different sizes and positions in an aluminum beam, under a set of combined loads under dynamic conditions. Strains were measured at several points by using Fiber Bragg Gratings.

WeAT5 KL Room
Structural Simulation (Regular Session)

Chair: Saravanos, Dimitris	Univ. of Patras
Co-Chair: Waeytens, Julien	Univ. Paris-Est, IFSTTAR

10:30-10:50 WeAT5.1

A Time-Domain High-Order Spectral Finite Element for the Simulation of Symmetric *and Anti-Symmetric Guided Waves in Laminated Composite Strips with Active Piezoelectric Sensors*

Rekatsinas, Christoforos	Univ. of Patras
Nastos, Christos	Univ. of Patras
Theodosiou, Theodosis	Univ. of Patras
Saravanos, Dimitris	Univ. of Patras

A new time domain spectral finite element is developed for improving the efficiency of numerical simulations of guided waves in laminated composite strips. The finite element relies on a new generalized laminate mechanics model formulated to represent symmetric and anti-symmetric Lamb waves. The laminate mechanics incorporate third-order polynomial terms for the approximation of axial and transverse displacement fields through the thickness, and consider the displacements of the upper and lower surfaces as degrees of freedom. Based on the resultant governing equations of the laminate section, a new finite element with 8 nodal degrees of freedom is formulated, with its nodes collocated with Gauss-Lobatto-Legendre integration points. Stiffness and mass matrices are assembled and the transient response is predicted with explicit central differences time integration. The transient response of an orthotropic composite strip exited by a 5-count Gaussian pulse is investigated. Results are validated against a semi-analytical solution Numerical results exhibit substantial improvement in the convergence and accuracy of the introduced element regarding the prediction of symmetric and anti-symmetric wave propagation.

10:50-11:10 WeAT5.2

Damage Assessment of Composite Plate Structures with Uncertainty

M., Chandrashekhar	Indian Inst. of Science
Ganguli, Ranjan	Indian Inst. of Science

Uncertainties associated with a structural model and measured vibration data may lead to unreliable damage detection. Two similar structures made from composite materials can display very different dynamic behavior due to large uncertainties associated with composite material properties. In this paper, we show that material uncertainties in composite structures cause considerable problem in damage assessment which can be alleviated by using a fuzzy logic-based approach for damage detection. A recently proposed robust Fuzzy Logic System (FLS) with sliding window defuzzifier is used for delamination damage detection in composite plate type structures. The FLS is designed using variations in modal frequencies due to randomness in material properties. Probabilistic analysis is performed using Monte Carlo Simulation (MCS) on a composite plate finite element model. A recently developed Reddy type C0 shear deformable locking free refined composite plate element is employed in the numerical simulations to alleviate modeling uncertainty. It is demonstrated that the FLS shows excellent robustness at very high levels of randomness in input data.

11:10-11:30 WeAT5.3

Detection of Structural Modifications of an Euler-Bernoulli Beam through Modal Parameters *Variation*

Cumunel, Gwendal	Lab. NAVIER (ENPC / IFSTTAR / CNRS)
Le, Thi-Thu-Ha	Lab. Navier (ENPC / IFSTTAR / CNRS)
Point, Nelly	Lab. Navier (ENPC / IFSTTAR / CNRS),
Argoul, Pierre	Lab. Navier (ENPC / IFSTTAR / CNRS)

This research deals with a class of identification problems mainly concerning the localization of structural modifications in elastic beams with or without tension by dynamic analysis. Variations of tension, mass density and/or bending stiffness induce variations of the beam natural frequencies. For each natural frequency, a first order estimate of this modification is then established. It depends on the variations of tension, density and bending stiffness and also on the corresponding mode shape and curvature mode shape of the initial state. When the variations of density and flexural rigidity are localized along the beam (as for a notch in a beam or a crack in a cable), this estimate is used to propose a procedure permitting detection, localization and, eventually, quantification of these variations knowing the modifications of the first natural frequencies and the variation of tension. This procedure is applied successfully to experimental data found in literature concerning beams without tension and only bending stiffness modifications.

11:30-11:50 WeAT5.4

Assessment of a Model-*Update Method for Damage Characterization of Critical Structures*

Viana, Julio	Critical Materials, SA
Antunes, Paulo	Critical Materials, SA
Vaz, Ismael	Univ. of Minho
Guimaraes, Rui	Critical Materials, SA
Ferreira, Nelson	Critical Materials, SA
Dias, Gustavo	Critical Materials, SA

A damage characterization approach is developed adopting a model-update method. A derivative-free global optimization based on the Particle Swarm method was adopted, combining data computed from a FEM model and experimental signals retrieved from accelerometers in order to estimate degradation of the material´s properties. The method is able of simultaneously locating, determining the type of damage (changes on the stiffness, mass, damping) and outputting its severity (size and variation in material properties) using a vibrational approach. The procedure is fully automated (using pre/post-processing operations, scripting capabilities, data exchange) and implemented into PRODDIA® - Structural Systems Health Management platform. The proposed approach results in a closed loop optimization process driven by an objective function that considers appropriate damage metrics. The damage assessment capability is validated in a sensorised CFRP plate, 1x1 m and 3 mm thick, with two Z-shape stringers. A data acquisition system with low number of sensors (4 accelerometers, 1 temperature sensor) is deployed. Several experimental signals are obtained by simulating damage through the placement of a mass at 25 different locations on the plate. Different damages metrics are also evaluated. The proposed approach is able of characterizing damage in complex composite plates at different locations.

11:50-12:10 WeAT5.5

Assessing Cracking Characteristics of Concrete Structures by Distributed Optical Fiber *and Non-Linear Finite Element Modelling*

Rodriguez Gutierrez, Gerardo — Tech. Univ. of Catalonia, UPC, Barcelona
Casas Rius, Joan Ramon — Tech. Univ. of Catalonia
Villalba Herreo, Sergi — Tech. Univ. of Catalonia

This work shows how the experimental strain data obtained with an OBR (Optical Backscattered Reflectometer) measuring system can be used to locate cracking before being visually observable and also to estimate the crack width for different levels of load. The method is checked in a test up to failure carried out on a reinforced concrete slab, where the deployement of other standard sensors allowed to validate the results from the OBR system. The results are also compared with those coming from a 2D non-linear finite element model, showing also a good agreement.

WeAT6 Room 200
Acoustic Emission (Regular Session)

Chair: Cawley, Peter — Imperial Coll.
Co-Chair: Le Duff, Alain — ESEO

10:30-10:50 WeAT6.1

Structural Health Monitoring of a Smart Composite Bridge Using Guided Waves and Acoustic Emission Techniques

Terrien, Nicolas — CETIM
Dahmene, Fethi — Ec.
Zejli, Hasnae — Alten
Drissi-Habti, Monssef — IFSTTAR

This paper is dealing with the development of a structural health monitoring (SHM) system implemented on a composite footbridge during the regional project “Pays de la Loire” called DECID2. The SHM system was made out of complementary techniques: strain sensors based on optical fibers (out of concern in this present work) and ultrasonic techniques that are presented in this document. Due to the huge size of the composite bridge (20 m * 3 m), only its most critical areas are monitored as the assembling parts and the most solicited areas. To access the structural integrity of the footbridge, two complementary monitoring strategies were presented in this paper: a real-time acoustic emission monitoring system to detect fibre breaks and a monitoring system using guided waves to evaluate the resin degradation. Both of these SHM systems use the same miniature ultrasonic patches that are used alternately as acoustic emission sensors and as ultrasonic guided waves actuators. The first step of this work was to develop these patches, and then to set up each monitoring systems and characterize their damage sensitivity. Finally, two composite footbridges were built at the EMC2 Technocampus and IFSTTAR in France to serve as demonstrators.

10:50-11:10 WeAT6.2

A New Approach for Fatigue Crack Length Estimation Using the Acoustic Emission Technique *in Structural Health Monitoring Applications*

Gagar, Daniel — Cranfield Univ.
Foote, Peter — Cranfield Univ.
Irving, Phil — Cranfield Univ.

The Acoustic Emission (AE) technique has been shown as capable in detecting and locating fatigue crack damage in metallic structures. However there are significantly fewer studies investigating its potential for fatigue crack length estimation. Information on the extent of crack growth would enable prediction of the remaining useful life of a component using well established fracture mechanics principles. This would improve the prospects of AE for use in structural health monitoring applications where detection and monitoring of crack lengths is required. A new approach for deducing absolute crack length has been developed based on correlations between AE signals generated during fatigue crack growth and corresponding cyclic loads. An empirical model to generate crack length was derived using AE data generated during fatigue crack growth tests in 2 mm thick SEN aluminium 2014 T6 specimens subject to a tensile stress range of 52 MPa and an R ratio of 0.1. The model was validated using AE data generated in separate tests performed with a stress range of 27 MPa. The results showed that predictions of crack lengths over a range of 10 mm to 80 mm can be obtained with the mean of the normalised absolute errors ranging between 0.28 and 0.4.

11:10-11:30 WeAT6.3

Topological Imaging of Defects in Anisotropic Plates

Rodriguez, Samuel — Univ. Bordeaux, I2M, UMR 5295, F-33400 Talence, France CNRS, I2M
Castaings, Michel — Univ. Bordeaux, I2M, UMR 5295, F-33400 Talence, France CNRS, I2M
Deschamps, Marc — Univ. Bordeaux, I2M, UMR 5295, F-33400 Talence, France CNRS, I2M
Ducasse, Eric — Univ. Bordeaux, I2M, UMR 5295, F-33400 Talence, France CNRS, I2M

Topological imaging is an emerging ultrasonic imaging method based on two computations performed for the so-called reference medium. The reference medium should correspond as close as possible to the experimental medium in the absence of defects. The topological image will then highlight all the differences between the investigated experimental medium and the reference medium. The presented work aims at applying this method to the detection, location and imaging of defects in composite plates. If the propagation can be properly simulated and the associated experience performed, the defects of the medium can be imaged whatever the complexity of the propagation process. Wave propagation in composite plates is here computed considering an equivalent homogeneous anisotropic medium in the frequency regime. First a numerical experiment is presented. It consists in the investigation of a strongly anisotropic medium using two different wave modes simultaneously. Three defects are accurately located and imaged in the medium. Secondly, a real experiment is investigating an impacted quadratic composite plate is investigated with a single guided mode. Two impacts are well detected and located.

11:30-11:50 WeAT6.4

Acoustic Emission Health *Monitoring of Steel Bridges*

Pahlavan, Pooria Lotfollah — TNO
Paulissen, Joep — TNO
Pijpers, Richard — TNO
Hakkesteegt, Henk — TNO
Jansen, Rob — TNO

Despite extensive developments in the field of Acoustic Emission

(AE) for monitoring fatigue cracks in steel structures, the implementation of AE systems for large-scale bridges is hindered by limitations associated with instrumentation costs and signal processing complexities. This paper sheds light on some of the most important challenges in the utilization of AE systems for steel bridge decks. These challenges are mainly related to the multi-modal character of guided waves, and the expensive installation of AE instrumentation. A quasi-beamforming solution which alleviates the above-mentioned challenges was introduced and successfully evaluated on a real-scale bridge segment in a laboratory environment. The solution was also implemented on a real bridge structure, demonstrating the benefit of the proposed configuration where the transducers are closely spaced for reduced installation costs.

11:50-12:10 WeAT6.5

Quantitative Damage Assessment of Hybrid Composite Wind Turbine Blades by Energy *Based Acoustic Emission Source Location*

Yoon, Dong-Jin — Korea Res. Inst. of Standards and Science

Han, Byeong-Hee — Korea Res. Inst. of Standards and Science

Acoustic emission technology was applied to assess the damage in the wind turbine blade. It was tried to apply a new source location method, which has a developed algorithm ourself with energy contour mapping concept. Firstly, we acquired energy based contour map database for tested blade section. And then, we measured the activities and the intensity of each arrival signals for several types of damage sources. That is, this study aims to locate and evaluate the damages such as internal damages or foreign impacts etc.. In this study, we focused to enhance a source location method with energy contour map which is developed already, and to develop a new damage index for more clear damage identification. For damage indexing, we found the correlation between corresponding energy and distance from source. Then, after calculating the location of damage source, we can doing more quantitative assessment using pre-acquired damage indexing. Consequently, the applicability of new source location method was confirmed by comparison of the result of source location and experimental damage location. From several experimental results, new suggested method of damage index identification showed very good performance for assessment of damages in the hybrid composites structures.

12:10-12:30 WeAT6.6

Acoustic Emission-Based Identification and Classification of Frictional Wear of Metallic Surfaces

Baccar, Dorra — Univ. of Duisburg-Essen

Schiffer, Sandra — Chair of Dynamics and Control, Univ. of Duisburg-Essen

Söffker, Dirk — Univ. of Duisburg-Essen

Diagnostic and monitoring approaches, able to detect and classify the damage and wear process, are becoming of increasing importance. Especially for friction wear-related phenomena which are difficult to measure directly during the operation and their diagnostic has been usually restrained to offline examinations. Permanent contact and repetitive sliding motions between two surfaces lead to material changes. Due to different wear mechanisms, sudden structural changes appear, emitting energy in form of elastic waves known as Acoustic Emission (AE). Therefore, a correlation between the emitted AE and damage level can give important information about the process state and the related knowledge can be used for automated supervision. This contribution introduces an advanced method for wear states identification and classification by means of AE technique and fuzzy-based multi-class classification approach. Compared to the previous publications, here the sequential effect of the motion trajectory is investigated. To establish a relationship between wear mechanism and AE signals, frequency-based feature selection using Continuous Wavelet Transform (CWT) was performed. Five wear process stages were detected during experiments. Results show that the behavior of individual frequency components changes when the wear-related effect changes. Using the CWT transformed signals, statespecific pattern are generated to classify signal features related to specific states. Results show that the introduced method can be used as an online monitoring method for material detection and characterization.

WeBT1 — Auditorium

Fiber Optics I (Regular Session)

Chair: Tur, Moshe — Tel-Aviv Univ.

Co-Chair: Chapeleau, Xavier — IFSTTAR

14:00-14:20 WeBT1.1

Identification of Damage Types in Carbon Fiber Reinforced Plastic Laminates by a Novel *Optical Fiber Acoustic Emission Sensor*

Yu, Fengming — The Univ. of Tokyo

Wu, Qi — Univ. of Tokyo

Okabe, Yoji — Univ. of Tokyo

Kobayashi, Satoshi — Tokyo Metropolitan Univ.

Saito, Kazuya — Univ. of Tokyo

In this research, phase-shifted FBG (PS-FBG) sensor was employed to practical AE detection for carbon fiber reinforced plastic (CFRP) composite laminate. Firstly, we evaluated the characteristics of AE signals detected by this kind of sensor. Secondly, through the experiment and simulation concerning AE source orientation, quantitative information about the standard for discriminating the AE signals due to transverse cracks and delaminations was obtained. Finally, according to the standard, we identified the occurrence of those two damage types successfully in the actual AE detection under three point bending and tensile test.

14:20-14:40 WeBT1.2

Adhesive Disbond Monitoring *with Microstructured Optical Fiber Bragg Grating* Sensors

Sulejmani, Sanne — Vrije Univ. Brussel - B-PHOT Team

Sonnenfeld, Camille — Vrije Univ. Brussel - Brussels Photonics Team

Geernaert, Thomas — Vrije Univ. Brussel - Brussels Photonics Team

Luyckx, Geert — Univ. Gent - Department of Material Science and Engineeri

Mergo, Pawel — Marie Curie-Sklodowska Univ. - Department of Optical Fiber

Urbanczyk, Waclaw — Wroclaw Univ. of Tech. - Inst. of Physics

Chah, Karima — Univ. of Mons - Electromagnetism and Telecom Department

Thienpont, Hugo — Vrije Univ. Brussel - Brussels Photonics Team

Berghmans, Francis — Vrije Univ. Brussel

We present a sensing system that is based on a combination of 3 optical fiber sensors which are non-intrusively integrated in an adhesive layer, and that allows detecting the initiation and monitoring the growth of disbonds. The fiber sensors consist of fiber Bragg gratings fabricated in a dedicated microstructured optical fiber and feature an enhanced response to shear stress. The change in response of the sensors to tensile loading of the joint is used as a measure for the shear stress redistribution in the bond layer caused by disbonding. With our technique we can detect the initiation of disbonds as short as 100 μ m, which corresponds to 0.4% of the overlap length. This is the first time, to the best of our knowledge, that an adhesive disbond on-line monitoring system is presented that can quantitatively evaluate even very short disbonds.

14:40-15:00 WeBT1.3

Structural Health Monitoring of Superconducting Magnets at CERN Using Fiber Bragg Grating *Sensors*

Chiuchiolo, Antonella — Optoelectronic Div. – Department of Engineering,

Univ.

Bajko, Marta — European Organization for Nuclear Res. CERN, Genève

Perez, Juan Carlos — European Organization for Nuclear Res. CERN, Genève

Guinchard, Michael — European Organization for Nuclear Res. CERN, Genève

Consales, Marco — Univ. of Sannio

Giordano, Michele — Inst. for Composite and Biomedical Materials, CNR

Breglio, Giovanni — Univ. of Naples Federico II

Bajas, Hugues — CERN European Organization for Nuclear Res.

Cusano, Andrea — Univ. of Sannio

The use of Fiber Bragg Grating sensors is becoming particularly challenging for monitoring different parameters in extreme operative conditions such as ultra-low temperatures, high electromagnetic fields and strong mechanical stresses. This work reports the use of the FBG for a new generation of accelerator magnets with the goal to develop an adequate sensing technology able to provide complementary or alternative information to the conventional strain gauges through the whole service life of the magnet. The study is focused on the mechanical performances of the magnet structure, which has to preserve the sensitive coils from any damage during the entire magnet fabrication process preventing even microscopic movements of the winding that can eventually initiate a transition from superconducting to normal conducting state of the material used (called in the specific literature as "quench"). The FBGs have been glued on the aluminium structure of two magnets prototypes by using an adhesive suitable for cryogenic temperature. The feasibility of the bonding procedure for bare sensors at 4.2 K leads to the validation of the final integration of the FBGs for the structural monitoring of the magnet during the assembly and cool down at 77 K.

15:00-15:20 WeBT1.4

Simultaneous Damage Detection and Deflection Measurement of Morphing Wing Structures by Fiber Optic Sensing *System*

Djinovic, Zoran — Austrian Center for Medical Innovation and Tech. Gmbh

Scheerer, Michael — Aerospace & Advanced Composites GmbH

Tomic, Milos — Univ. of Belgrade

Stojkovic, Marijana — Integrated Microsystems Austria

Schueller, Martin — Fraunhofer ENAS

In this paper we present results of investigation of simultaneous damage detection and deflection measurement of morphing CFRP honeycomb structure by fiber optic sensing system developed in the frame of EU-FP7 project "Fiber Optic System for Deflection and Damage Detection (FOS3D)". The system is based on low- and high-coherence interferometry performed as "all-in-fiber" sensing configuration. Raw signals have been on- and off-line processed by "arctang" algorithm. Deflection data, expressed as phase angle change of the interferometer, have been simultaneously acquired with the reference data. A linear relationship between the fiber-optic sensing and reference technique is obtained. Slope of the linear fit line of 8 rad/mm denotes sensitivity of this sensor. Noise floor of about ±70µrad/√Hz determines the lowest measurable wing deflection of about 6µm. Damage events have been simulated by pencil break and hammer impact with- and without damages over the CFRP honeycomb structure. Raw fiber-optic interference and reference acoustic emission sensor signals have been simultaneously acquired. Figure of merit is given as probability of detection (POD) and localization of simulated impact event. All events can be detected and located with a POD of more than 98% within the localization error of around 15 mm.

15:20-15:40 WeBT1.5

Extended Hierarchical Fiber-Optic-Based System for Sensing-Healing of Composite Delamination

Minakuchi, Shu — The Univ. of Tokyo

Sun, Denghao — The Univ. of Tokyo

Takeda, Nobuo — The Univ. of Tokyo

This study demonstrates the first autonomous sensing-healing system applicable to large-scale composite structures. Microvascular self-healing concept is combined with our hierarchical fiber-optic-based sensing system. The hierarchical system is a fluid distribution system with a local pressure monitoring function, and thus the combined system offers the ability to sense and heal composite delamination, and also to self-diagnose the system condition. This study begins by describing the basic concept of the combined system. The feasibility of the system is then evaluated through delamination infiltration tests. The system is finally validated in a plate specimen.

15:40-16:00 WeBT1.6

Performance Evaluation on Core Wire of Smart Strands for PSC Structures

Kim, Sung Tae — Korea Inst. of Construction Tech.

The Korea Institute of Construction Technology is undertaking the development of the smart strand for PSC structures enabling to measure the prestress force all along its service life since the completion of construction. Apart from such measurement function, the smart strand shall at first satisfy the structural performance required as a tendon. Therefore, research has been implemented so as to embed the sensor within the small diameter of the core wire by using optical fibre sensor. To that goal, new materials were devised for the core wire to avoid the method cutting and machining the conventional steel core wire. The performances of the core wire fabricated by this new approach as well as its fabricability and structural performance were examined through tensile test to derive the optimal fabrication method of the smart strand. The tests conducted on the core wire fabricated by the so-selected fabrication method and embedded with the optical fibre sensor revealed that the resulting strain measurement range was significantly larger than the range complying with the design prestress force of the conventional steel strand. The core wire for the smart strand under development was seen to provide sufficient measurement performance as well as to satisfy the required structural performance. Accordingly, the newly developed core wire can be conveniently applied in the conventional steel strand.

WeBT2 GH Room

Probabilistic SHM 2 (Invited Session)

Chair: Glisic, Branko — Princeton Univ.

Co-Chair: Lanata, Francesca — Ec. Supérieure du Bois

14:00-14:20 WeBT2.1

Damage Identification *of Nonlinear Structure with Unknown Excitations Using Quadratic Sum Square Error with Ar Model (I)*

Huang, Hongwei — Tongji Univ.

The ability to accurately identify structural parameters, either on-line or almost on-line, based on vibration data measured from sensors, is essential for the structural health monitoring system. The problem is quite challenging, in particular when the external excitations are not completely measured and when the structural system is nonlinear. In practical applications, external excitations (inputs), such as seismic excitations, wind loads, traffic loads, etc., may not be measured or may not be measurable, and the structure may not always be linear. In this paper, a newly proposed parametric identification method, referred to as the quadratic sum-squares error with AR model (QSSE-AR), is used for estimating structural parameters of a nonlinear elastic structure and a nonlinear hysteretic structure. In this approach, external excitations and some structural responses may not be measured. The accuracy and effectiveness of the proposed approach will be demonstrated by numerical simulations without the measurement of external excitations. The simulation results indicate that the proposed damage detection technique is capable of identifying structural parameters, as well as predicting the unknown external excitations.

14:20-14:40 WeBT2.2

Algorithm Fusion in Novelty *Detection (I)*

Bernal, Dionisio — Northeastern Univ.

Algorithm fusion has received significant attention in the machine learning community in supervised learning mode but it appears little has been done at this point in a novelty detection framework. This paper examines the merit of a fusion strategy wherein metrics from multiple algorithms are treated as entries of a vector whose probability density is subsequently estimated and used for detection. In the present paper the framework is investigated using two algorithms: 1) a robust version of a whiteness test on Kalman filter innovations and 2) a robust version of a scheme that operates with residuals obtained from an orthogonality test. The density estimation part of the process is replaced by the Kernel PCA algorithm which provides a decision boundary without having explicit density estimates. The fused scheme is implemented in a change detection format and is show to provide notable improvements over the use of either algorithm independently.

14:40-15:00 — WeBT2.3

Stochastic Evaluation of Carbonation Depth and Predictions Updating (I)

Decatoire, Rodrigue	PHIMECA Engineering
De Larrard, Thomas	Lab. Durabilité et Matériaux de Constructions, UPS-INSA T
Yalamas, Thierry	PHIMECA
Schoefs, Franck	Univ. of Nantes
Elachachi, Sidi Mohammed	Univ. of Bordeaux

In the scope of the inspection plan optimization of civil engineering structures subject to carbonation (carbon dioxyde penetration), several analytical models help to predict the degradation evolution. The degradation model used here is coming from the european project DuraCrete. This paper investigates its predictive capability, using the a priori of its inputs given in the litterature, but also after updating its predictions based on simulated measures with a finite element code.

15:00-15:20 — WeBT2.4

Probabilistic Damage Detection *Based on Large Area Electronics Sensing Sheets (I)*

Yao, Yao	Princeton Univ.
Glisic, Branko	Princeton Univ.

Reliable early-stage damage detection and characterization requires continuous sensing over large areas of structure. The limitations with current sensing technologies lies in the fact that they either have high cost and insufficient spatial resolution, or rely on complex algorithms that are challenged by varying environmental and loading conditions. This paper addresses the need for direct sensing where anomalies are sensed at close proximity through a dense array of sensors, and proposes one approach for sensor network design. This approach is directly applicable to innovative sensing sheet based on large area electronics (LAE), which enables practical implementation of dense arrays of sensors. However, although the sensors are densely spaced in the sensing sheet, there are still some non-instrumented spaces between them and these spaces are not sensitive to damage. In this research, a probabilistic approach based on Monte Carlo (MC) simulations is researched to determine the probability that damage of certain size that occurs within the area covered by the sensing sheet can be detected with a given sensor network. Based on these Probability of Detection (POD) functions, it was possible to assess the reliability of sensing sheets for crack detection and to establish general principles for the design of sensing sheets.

15:20-15:40 — WeBT2.5

Bayesian Updating of Probabilistic Time-Dependent Fatigue Model: Application to Jacket *Foundations of Wind Turbines (I)*

Schoefs, Franck	Univ. of Nantes
François, Marc	Faculté des Sciences de Nantes, Lab. GeM
Rocher, Benjamin	Univ. of Nantes
Salou, Arnaud	TX France Solutions SAS

The metal foundations of offshore wind turbines are highly sensitive to the fatigue phenomenon. To date, current methods of fatigue design proposed in the regulations are an obstacle to the structural optimization and to the consideration of hazards. In this context, we propose an incremental two scales model of damage in order to follow the time evolution of the damage parameter. This time notion is important to the update of the model parameters using records from the Structural Health Monitoring. In this article, we focus on updating the parameters of the damage model using experimental data and the method of Bayesian updating based on an MCMC algorithm.

15:40-16:00 — WeBT2.6

Deterioration Forecasting in Flexible Pavements Due to Floods and Snow Storms *(I)*

Shahini Shamsabadi, Salar	Northeastern Univ.
Hashemi Tari, Yasamin Sadat	Northeastern Univ.
Birken, Ralf	Northeastern Univ.
Wang, Ming	Northeastern Univ.

Roadway agencies and state DOTs utilize Pavement Management Systems (PMS) to implement cost-effective maintenance strategies. A reliable yet easily applicable model for deterioration process of pavements is an integral part of any Pavement Management System. As pavement condition grows to be one of the crucial problems facing our nation, the reliability of these deterioration prediction models becomes more important. While numerous endevours have been made to capture the effect of the environment, load and pavement's structure on pavement failures, only few have realized the impact of severe events such as Snow Storms and Floods on road infrastructures. First, this impact was quantified using Long Term Pavement Performance (LTPP) and National Oceanic and Atmospheric Administration (NOAA) databases with a dependable natural deterioration model. Then, a regression-based statistical approach has been undertaken to model the effect of snow storms and floods on pavement serviceablilites based on the severity of the events and condition of the pavement prior to these event. Final models rendered more than 90% correlation with the quantified impact values of snow storms and floods.

WeBT3 — Room I

Sensors IV (Regular Session)

Chair: Casciati, Fabio	Univ. of Pavia
Co-Chair: Drissi-Habti, Monssef	IFSTTAR

14:00-14:20 — WeBT3.1

A Decentralized Approach towards Autonomous Fault Detection in Wireless Structural *Health Monitoring Systems*

Smarsly, Kay	Bauhaus-Univ. Weimar
Petryna, Yuri	Tech. Univ. Berlin

Sensor faults in wireless structural health monitoring (SHM) systems may reduce the monitoring quality and, if remaining undetected, might cause substantial economic loss due to inaccurate or missing sensor data required for structural assessment and life-cycle management of the monitored structure. Usually, fault detection in sensor networks is achieved through a redundant deployment of sensors and further hardware components ("physical redundancy"), which involves considerable penalties in cost and maintainability. Overcoming these drawbacks, in this study the information inherent in the SHM system and the known relationships between the sensors are used for fault detection without the need for additional sensors ("analytical redundancy"). Furthermore, the analytical redundancy approach is implemented in a fully decentralized manner: Partial models of the SHM system, being embedded directly into the wireless sensor nodes, enable each sensor node to autonomously detect sensor faults in real time while efficiently using the limited computing resources.

14:20-14:40 — WeBT3.2

Development of a Reliable Wireless GNSS Sensors Network for

Asset Tracking in Emergency Management

Casciati, Fabio	Univ. of Pavia
Casciati, Sara	Univ. of Catania
Chen, Zhicong	Univ. of Pavia
Faravelli, Lucia	Univ. of Pavia
Vece, Michele	Univ. of Pavia

To localize, track and trace critical assets and first responders for remote emergency management are primary needs after a catastrophic event. In disaster areas, however, it is likely that the existing terrestrial communication network and the power grid will be cut off. Therefore, a reliable asset tracking system should be independent of those terrestrial networks, especially in terms of communication. This paper explores the development of a reliable low power local wireless communication network for portable GNSS-based positioning sensors and a centralized recording unit, as part of the independent communication network. The wireless sensors network is based on the ZigBee standard protocol and the IEEE802.15.4 compliant transceiver CC2530 from Texas Instrument. The network features low cost, low power, mesh topology, high reliability and high security.

14:40-15:00 WeBT3.3

An On-Line Wireless Impact *Monitoring System for Large Scale Composite Structures*

Mei, Hanfei	Nanjing Univ. of Aeronautics and Astronautics
Yuan, Shenfang	Nanjing Univ. of Aeronautics and Astronautics
Qiu, Lei	Nanjing Univ. of Aeronautics and Astronautics
Ren, Yuanqiang	Nanjing Univ. of Aeronautics and Astronautics

One of the major concerns in the whole lifetime of aircraft composite structures is their susceptibility to impact damage. Aiming at the impact monitoring of large scale composite structures, this paper puts forward an on-line wireless impact monitoring system based on a kind of wireless digital impact monitor developed. Different from traditional processing methods, the new one is fulfilled in a digital way by turning the outputs of PZT sensors directly into digital queues and localizing the impact occurring sub-region based on the digital queues obtained. Meanwhile, validation experiments are implemented on an unmanned aerial vehicle (UAV) composite wing. The results show promising performance of the proposed system for on-line structural health monitoring (SHM) applications on large scale aircraft composite structures.

15:00-15:20 WeBT3.4

Wireless Passive Sensing *Platform Based on RFID Technology for SHM Purposes*

Lisowski, Mateusz	AGH Univ. of Science and Technology, Kraków
Gonek, Przemyslaw	AGH Univ. of Science and Technology, Kraków
Uhl, Tadeusz	AGH Univ. of Science and Technology, Kraków
Staszewski, Wieslaw J.	Department of Robotics and Mechatronics, AGH Univ. of Science and Technology, Kraków

Recently there are many attempts to use wireless communication and its capabilities in data acquisition from sensors, in SHM application. Beside typical active Wireless Sensor Networks (WSN), there are also attempts to develop sensor networks with passive nodes, using the advantages, provided by Radio Frequency Identification (RFID) technology. The main benefits of this solution are utilization of energy of the communication signal from the reader antenna for sensor circuit powering and possibility to place sensors in any places, including hard available and moving locations. Development of such sensing platform based on M24LR64E-R RFID chip working in HF band, suitable for working with different types of sensor - strain gauges, accelerometers, pressure sensors is presented. It is designed especially for monitoring helicopter's rotor blades. Entire energy needed to power the microcontroller, RFID chip, sensor and other circuit elements is supplied by the reader wirelessly, by using magnetic coupling between the reader and the sensing platform antenna. The paper presents a concept of the sensing platform, its first prototypes and investigation related to communication between the platform and the reader, as well as first experimental studies related to using this platform in different applications.

15:20-15:40 WeBT3.5

Wireless and Autonomous *Sensor for Strut Load Monitoring*

Bense, William	SNECMA
Florent, Nicolas	SNECMA
Frutos, Jean-Renaud	SILMACH
Hebrard, Yoann	SKF-AEROSPACE

In case of hard landings, engine mounts or landing gears can be damaged and must be inspected as the severity of landing is not well estimated based on aircraft inertial system and models. Such inspections can be time consuming and generate delays, cancellations and inappropriate maintenance actions that could be avoided if a sensor is able to measure the accurate load level sustained by the equipment. The aim of the project is to develop an autonomous wireless sensor with local intelligence and storage capacity that can be downloaded upon request. A micro-mechanic MEMS sensor (chronoMEMS), using cogwheels, has been implemented: when a predefined level of load is reached, the sensor increments a counter. To validate the discrete information given by this sensor, strain gauges have been added. Interrogating the sensor must be easy and fast so that a wireless interrogating system based on RFID has been chosen. An Energy Harvesting system transforming Thermal and Vibration into electric power has been developed to supply the strain gauges and the memory (chronoMEMS does not require power). Lab tests have been done on thermal and vibration energy harvesters. RFID transmission in the engine area is yet under studies and tests. The next step is to develop a prototype of a full sensor integrating all these parts working together in the engine harsh environment. This technology could be used to monitor any kind of load charge for which real time surveillance is not necessary. Snecma (Safran Group) is the leader of this project allied to SilMach and SKF Aerospace. SilMach develops the chronoMEMS load sensor. SKF Aerospace is in charge of strain gauges, energy harvesting, RFID modules and complete system assembly.

15:40-16:00 WeBT3.6

Wireless Sensors *Embedded in Concrete*

Abbadi, Amal	Univ. LILLE1

Efficient embedded antennas are needed for future wireless structural health monitoring. The properties of patch antennas with concrete are investigated at 860MHz. Simulations for different cases (different concrete permittivity and tangent loss) with and without the presence of steel reinforcements are performed using ANSYS HFSS software for 3-D full-wave electromagnetic field simulation and some experimental results are presented.

WeBT4 Room J

Pattern Recognition II (Regular Session)

Chair: Uhl, Tadeusz	AGH Univ. of Science and Technology, Kraków
Co-Chair: Argoul, Pierre	Lab. Navier (ENPC / IFSTTAR / CNRS)

14:00-14:20 WeBT4.1

Fuzzy Similarity Classifier As Damage Index: Temperature Effect and Compensation

Gharibnezhad, Fahit	Tech. Univ. of Catalunya
Mujica, Luis Eduardo	Tech. Univ. of Catalonia
Rodellar, Jose	Tech. Univ. of Catalonia
Todd, Michael	Univ. of California, San Diego

The effects of ambient operational temperature variability on the measured dynamics response of structures have been addressed in several studies. It is intuitive that temperature variation may change the material/geometric properties or boundary conditions of a structure and therefore may affect the damage detection performance. Then we consider the ability of a Fuzzy similarity classifier as a feature when the temperature is changing; it will be shown that temperature change might have more significant effect rather than the simulated damage on this feature, which leads to false positive decisions. Therefore, it is vital to compensate the effect of temperature to achieve a desirable result. To do this, the temperature effect is compensated and it is shown the compensation increases the performance of damage detection using the Fuzzy similarity index. To support claims mentioned above, this work involves experiments with composite plate equipped with PZT transducers. To simulate the effect of temperature the specimen is subjected to temperature change between -25C and 60C.

14:20-14:40	WeBT4.2

Operational Modal Analysis and Wavelet Transformation for Damage Identification in Wind Turbine Blades

Ulriksen, Martin Dalgaard	Aalborg Univ.
Tcherniak, Dmitri	Bruel and Kjaer Sound and Vibration Measurement
Kirkegaard, Poul Henning	Aalborg Univ.
Damkilde, Lars	Aalborg Univ.

The presented study demonstrates an application of a previously proposed modal and wavelet analysis-based damage identification method to a wind turbine blade. A trailing edge debonding was introduced to a SSP 34m blade mounted on a test rig. Operational modal analysis (OMA) was conducted to obtain mode shapes for undamaged and damaged states of the blade. Subsequently, the mode shapes were analyzed with one-dimensional continuous wavelet transformations (CWTs) for damage identification. The basic idea of the method is that structural damage will introduce local mode shape irregularities which are captured in the CWT by significantly magnified transform coefficients, thus providing combined damage detection, localization, and size assessment. It was found that due to the nature of the proposed method, the value of the identification results highly depends on the number of employed measurement points. Since only a limited number of measurement points were utilized in the experiments, valid damage identification can only be obtained when employing high-frequency modes.

14:40-15:00	WeBT4.3

Data Driven Methodology Based *on Artificial Immune Systems for Damage Detection*

Anaya, Maribel	Univ. Pol. de Catalunya-BarcelonaTech
Tibaduiza, Diego A.	Univ. Santo Tomás
Pozo, Francesc	Univ. Pol. de Catalunya

Structural Health Monitoring is a growing area of interest given the benefits obtained from its use. This area includes different tasks in the damage identification process, the main important, is the damage detection since an early detection allows to avoid possible catastrophes in structures in service. Practical solutions require a big quantity of sensors and a robust system to process and obtain a reliable solution. In this sense, bio-inspired algorithms provide tools for an effective data analysis taking advantage of the developments provided by the nature by means of computational algorithms. As a contribution in this area, this paper presents a methodology for structural damage detection using a type of artificial intelligence that is called artificial immune system. The developed methodology includes the inspection of the structure by means of a distributed piezoelectric active sensor network at different actuation phases to define a baseline by each actuation phase using data from the structure when it is known as healthy. In a second step, same experiments are performed to the structure when its structural state is unknown to determine the presence of damage by using the developed artificial immune system. Results show that the proposed methodology allows to detect damages in the experimental setup.

15:00-15:20	WeBT4.4

Statistical Impact-Echo Analysis Based on Grassmann Manifold Learning: Its Preliminary *Results for Concrete Condition Assessment*

Ye, Jiaxing	National Inst. of Advanced Industrial Science and Tech.
Iwata, Masaya	The National Inst. of Advanced Industrial Science and Tech.
Takumi, Kobayashi	National Inst. of Advanced Industrial Science and Tech.
Murakawa, Masahiro	The National Inst. of Advanced Industrial Science and Tech.
Tetsuya, Higuchi	The National Inst. of Advanced Industrial Science and Tech.
Kubota, Yuichi	Shutoko Engineering Company Limited.
Yui, Toshiya	Shutoko Engineering Company
Mori, Kiyoshi	Shutoko Engineering Company

Impact-echo is one extensively applied non-destructive technique for flaw detection in concrete structures. In impact-echo test, surface motion generated by short-duration mechanical impact is investigated for structural condition assessment. This paper endeavours to formulate impact echo analysis by using novel statistical techniques, i.e. Grassmann manifold learning. Comparing to conventional impact-echo test, the proposed method presents several favourable properties: 1. Conventional impact-echo method mostly relies on frequency peak in echo spectrum; the proposed method characterizes rich temporal-spectral patterns in addition to the spectral peak. 2. Proposed method is performed over local area on concrete surface with integration of several consecutive echo responses, and thus produces more stable condition evaluation result comparing to point-wise impact-echo approach. 3. To cope with extracted echo feature, effective similarity metric on Grassmann manifold is employed, which favourably facilities condition-based assessment. To demonstrate the proposed method, we prepared concrete specimen with 2cm, 4cm and 6cm depth void inside and echo signal is captured through air-coupled sensor. Experimental result demonstrates the effectiveness of the proposed method, including accurate condition-based classification performance and high processing efficiency

15:20-15:40	WeBT4.5

Baseline-Based Detection *of Structural Changes: Drawbacks and an Alternative Approach*

Santos, João	Lab. Nacional de Engenharia Civil (LNEC)
Calado, Luís	IST, Univ. of Lisbon
Silveira, Paulo	Lab. Nacional de Engenharia Civil (LNEC)
Crémona, Christian	SÉTRA

The present paper addresses the detection of structural changes using statistical learning methods with the objective of exposing the major limitations of defining baselines in which structural systems must be assumed unchanged. For this purpose, the data acquired during approximately two years in a cable-stayed bridge is used and was chosen since it comprises unchanged structural responses as well as measurements comprising changes imposed by creep and shrinkage. In addition to the study of baseline-based detection, the present paper proposes a simple yet effective baseline-free approach relying auto-correlation functions. The effectiveness of the proposed strategy in clearly distinguishing between changed and unchanged structural responses was demonstrated using the same data set.

15:40-16:00	WeBT4.6

Adaptive Detection of Structural Changes Based on Unsupervised Learning and Moving *Time-Windows*

Santos, João	Lab. Nacional de Engenharia Civil (LNEC)
Calado, Luís	IST, Univ. of Lisbon
Orcesi, André	IFSTTAR

Crémona, Christian SÉTRA

The present paper addresses data-driven structural health monitoring to propose a real time strategy for adaptive structural assessment. The adaptive character is achieved using unsupervised discrimination machine-learning methods, widely known as clustering algorithms. Real-time capability is based on the definition of symbolic data, which allow describing large amounts of information without loss of related information. The efficiency of the proposed methodology is illustrated using an experimental case study in which structural changes were imposed to a suspended bridge during an extensive rehabilitation program.

WeBT5 KL Room
FEM Methods and SHM (Regular Session)

Chair: Stepinski, Tadeusz — AGH Univ. of Science and Technology, Kraków

Co-Chair: Ramasso, Emmanuel — Femto-st Inst.

14:00-14:20 WeBT5.1

Finite Element Model Updating *Using Spectral Density of Structural Response*

Pedram, Masoud — Amirkabir Univ. of Tech.
Esfandiari, Akbar — Amirkabir Univ. of Tech.
Shadan, Fariba — Amirkabir Univ. of Tech.

Finite element models (FEM) are used for estimation of the physical behavior of the structures may not accurately represent real structural responses because of inaccurate assumption of mechanical properties. For the achievement of a reliable model, the numerical responses must be adjusted to the experimental or real life responses through a proper model updating scheme. The Power Spectral Density (PSD) of the structural response subjected to a random input is opted in this research for derivation of sensitivity based model updating method. Taylor series expansion of the PSD is used to develop the sensitivity equations. Upon the consideration of all the terms of the expansion, it is possible to take advantage from an already developed method based on frequency response function (FRF). The proposed sensitivity equation has also eliminated the need for model reduction or expansion to deal with incomplete measurement which is required for most of updating methods relying on the modal data. The proposed method is examined using a numerical example to predict changes in stiffness parameters.

14:20-14:40 WeBT5.2

Finite Element Model-Based Structural *Health Monitoring (SHM) Systems for Composite Material under Fluid-Structure Interaction (FSI) Effect*

Ebna Hai, Bhuiyan Shameem Mahmood — Helmut Schmidt Univ. Univ. of the Federal Armed Force
Bause, Markus — Helmut Schmidt Univ. Univ. of the Federal Armed Force

Advanced composite materials such as Carbon Fibre Reinforced Polymers (CFRP) tend to be used in aerospace industry to keep the weight at its minimum and yet retain a great strength. CFRP have a strong, stiff fibres in a matrix. The resulting material is very strong as it has the best strength to weight ratio of all construction materials. However, aircraft structures such as wings can break due to Fluid-Structure Interaction (FSI) oscillations or material fatigue. Material inspection by piezoelectric induced ultrasonic waves is a relatively new and an intelligent technique to monitor the health of CFRP for a damage detection in the Non-Destructive Test (NDT). To design a Structural Health Monitoring (SHM) system, it is important to understand phenomenologically and quantitatively the wave propagation in CFRP and the influence of the geomaterial and mechanical properties of the structures. The principal aim of this research is to explore and understand the design and operation from safety and economic points of view. To accelerate the design of SHM systems, the FSI effect on the wave propagation has to be considered. This research will focus on the mathematical modeling and numerical analysis of Navier-Stokes, elastodynamics and elastic waves equations in the arbitrary Lagrangian-Eulerian (ALE) framework in order to determine the wave propagation in moving domains and optimum locations for sensors. Since analytical solutions are only available in special cases, the equations need to be solved by numerical methods. For the implementation we chose the finite element library package deal.ii and DOpElib due to its special characteristics.

14:40-15:00 WeBT5.3

Calibration of the Finite Element Model of a Twelve-Span Prestressed Concrete Bridge *Using Ambient Vibration Data*

Chen, Xinghua — Univ. of Auckland
Omenzetter, Piotr — Univ. of Aberdeen
Beskhyroun, Sherif — The Univ. of Auckland

The recently constructed Newmarket Viaduct in Auckland is a critical link in the New Zealand state highway network. Newmarket Viaduct is a 12-span, pre-cast, post-tensioned structure built using the balanced cantilever method. A continuous health monitoring system was designed and installed in the bridge. As a critical part in the SHM process, a baseline finite element (FE) model of Newmarket Viaduct was established. This paper describes the implementation of the FE model calibration using ambient vibration data. The initial model of the bridge was developed from the information provided in the design documentation, material testing data and site inspections. Two ambient vibration testing campaigns used some 60 wireless sensors in multiple setups to collect data to map with high density 3D mode shapes of the bridge. The output-only modal identification results obtained from the ambient vibration measurements of the bridge were used to update the FE bridge model. Different parameters of the model were calibrated using an automated procedure to improve the correlation between measured and calculated modal parameters. Careful attention was paid to the selection of the parameters to be modified during updating in order to ensure that the changes to the model were realistic and physically meaningful. The calibrated FE model reflecting the as-built structural condition and dynamic response mechanisms of Newmarket Viaduct will serve as a baseline model for assessment of structural health using continuous monitoring data.

15:00-15:20 WeBT5.4

Non-Destructive Determination of Serviceability and Load Bearing Capacity of Floor Slabs Using Dynamic Methods

Griessmann, Tanja — Leibniz Univ. Hannover
Penner, Nikolai — Leibniz Univ. Hannover
Schmoch, Arne — Leibniz Univ. Hannover
Rolfes, Raimund — Leibniz Univ. Hannover, Inst. of Structural Analysis

Due to dwindling resources and the aging of the building infrastructure, many future activities in civil engineering will focus on building conversions and refurbishments. The key prerequisite in this case is the verification of the serviceability and the load bearing capacity towards the certification authorities. In most of the practical cases the current loading status, the internal stresses and deformation under static loads are unknown. As a result, the immanent load reserves can't be exploited, which often leads to a very conservative approach or even to demolition. Refurbishing of existing slabs means dealing with many uncertainties. Material properties like densities or elasticity modules as well as geometries or boundary conditions are often unknown. For this reason there's a strong need to make efficient use of all information derived from measurements under static and dynamic test loads. In this context, in the field of modal analysis especially output-only methods utilizing only response measurements have proved particularly powerful. The big advantage is that the exciting forces need not to be measured. As a consequence high demands on signal analysis and the subsequent system identification are made. This paper describes a procedure to automatically update a numerical model by means of a prior identification of modal parameters with the Frequency Domain Decomposition (FDD). The modal parameters are used to calibrate the numerical model, which uses a-priori information on the construction. The updated model is subsequently capable to describe the dynamic behavior of the slab within the considered frequency range. The different steps of the methodology are shown using the example of a wooden beam ceiling.

15:20-15:40 WeBT5.5

Comparison of Infinite Element Models

Palacz, Magdalena — Gdansk Univ. of Tech.
Skarbek, Lukasz — Faculty of Electrical and Control Engineering, Gdańsk Univ.
Zak, Arkadiusz — Faculty of Electrical and Control Engineering, Gdańsk Univ.

The main objective of this paper is to show the comparison of two models of infinite absorbing layer with increasing damping in numerical investigations of elastic wave propagation in unbounded structures. This has been achieved by the Authors by a careful investigation of two different engineering structures characterised by gradually increasing geometrical and mathematical description complexities. The analysis included propagation of longitudinal elastic waves in a 1-D half-infinite isotropic rod, modelled according to the classical 1-mode theory of rods as well as propagation of coupled shear and flexural elastic waves in a 1-D half-infinite isotropic beam modelled according to the Timoshenko beam theory. The comparison of both models has been not only presented by the Authors, but also advantages and disadvantages of both of them have been discussed.

WeBT6 — Room 200
Advances in Monitoring with Ultrasonic Coda Waves (Invited Session)

Chair: Tournat, Vincent — CNRS, Univ. du Maine, LAUM
Co-Chair: Niederleithinger, Ernst — BAM

14:00-14:20 WeBT6.1

Monitoring of a Large Cracked Concrete Sample with Non-Linear Mixing of Ultrasonic Coda Waves (I)

Abraham, Odile — IFSTTAR
Zhang, Yuxiang — ISTerre
Chapeleau, Xavier — IFSTTAR
Durand, Olivier — IFSTTAR
Tournat, Vincent — CNRS, Univ. du Maine, LAUM

A high precision can be achieved with ultrasonic coda waves to monitor the mechanical properties of concrete material (~10-5 in relative). This high sensitivity can be used to detect damage initiation and to closely follow concrete mechanical properties evolution with time. This advantage is counterbalance by the influence of environmental conditions making reproducibility of any experiment in concrete a challenging issue especially when in situ measurements are performed. Indeed thermal and water gradients present in the thickness of the structures (several decimetres) cannot be controlled and must be compensated. In this paper a protocol to remove environmental bias is proposed. Furthermore, to follow the apparition of a tensile crack in a metric size structure, non-linear mixing of coda wave via frequency-swept pump waves is tested. It is shown that, when the crack is closed (by pre-stressing cables), it is still possible to detect its presence. The non-linearity of the cracked zone remains at a high level, comparable to the case when the crack was open.

14:20-14:40 WeBT6.2

Imaging of Early-Stage Cracking on Real-Size Concrete Structure from 4-Points Bending Test (I)

Zhang, Yuxiang — ISTerre
Larose, Eric — ISTerre, Univ. J. Fourier & CNRS
Planes, Thomas — Dept. of Civil and Environmental Engineering, Colorado School
Moreau, Gautier — EDF R&D
Rospars, Claude — Lab. Sols, Solides, Structures - Risques (3S - R) / EDF /

Traditional ultrasonic imaging techniques encounter difficulty on complexes material such as concrete, which is in part due the use of coherent waves in a very heterogeneous material. From this angle, technique called LOCADIFF has been developed for monitoring heterogeneous media using multiply scattered waves [1, 2]. We consider that modifications in the medium are equivalent to the presence of extra scatterers, which are characterized by their effective scattering cross-section σ. Within this view, LOCADIFF allows to locate the modification by measuring the spatio-temporal de-correlation of multiply scattered waves and by solving the corresponding inverse problem. Based on LOCADIFF, a newly developed imaging technique has been reported [3]. By mapping the intensity of modification on localized microstructure, the new technique is able to detect perturbations at multiple locations. Here we present the application of this new technique on a real-size 15 tons concrete structure for imaging early-stage cracking procedure issued from four point bending load, as part of the CEOS.fr project. Experimental results show that this technique can not only locate cracks that appeared simultaneously at multiple locations, but also detect them and observe their developments since an early-stage.

14:40-15:00 WeBT6.3

Coda Wave Interferometry Used to Localize Compressional Load Effects on a Concrete Specimen (I)

Niederleithinger, Ernst — BAM
Sens-Schönfelder, Christoph — GFZ Potsdam
Grothe, Sven — BAM
Wiggenhauser, Herbert — BAM

Coda Wave Interferometry (CWI), a method to evaluate subtle changes of elastic wave velocity in a medium, has been proven to be effective to detect small changes or ultrasonic velocity in concrete caused by load, temperature, moisture, damage or other means. While classical CWI is just able to determine velocity changes globally in relatively large areas between and around pairs of transmitters and receivers, several approaches have been proposed to identify the area affected by the changes more precisely. Most of them are based on the calculation of sensitivity kernels for de-correlation of signals measured at a specific state against a reference. Others follow simplified approaches. In a laboratory setup a concrete specimen of 1.5 by 1.5 by 0.5 m^3 was compressed at a certain point by a thread rod, screw nuts and 10 by 10 cm^2 load distribution plates including a load cell to measure the effective load. Maximum loads of 20 to 100 kN (more than one orders of magnitude below the compressive strength of the concrete) have been applied in 5 to 10 kN steps in various cycles. The specimen is equipped with 18 embedded ultrasonic broadband piezo transceivers (60 kHz central frequency). Ten of these receivers have been connected to a multiplexer and ultrasonic transmitting and receiving equipment in a way that allowed almost continuous two way measurements between all sensor pairs. Some of the sensors have been just a few cm away from the center of the load, some almost 1 m. Even simple ways to evaluate the data (e.g. crosscorrelation between signals at different load states) allowed pinpointing the load center at least approximately. A more detailed data evaluation either using CWI or even more one of the more sophisticated localization algorithms gave "sharper" results in terms of localization and a better correlation between load and velocity change/de-correlation. Referencing the results to the transmitter-receiver pair least affected by load change has led to further improvement. The results are used in upcoming monitoring systems for concrete structures.

15:00-15:20 WeBT6.4

Closed Crack Detection in Concrete with Coda Wave Non-Linear Modulation (I)

Hilloulin, Benoit — Ec. Centrale de Nantes
Zhang, Yuxiang — ISTerre
Abraham, Odile — IFSTTAR
Grondin, Frédéric — Ec. Centrale de Nantes
Loukili, Ahmed — Ec. Centrale de Nantes
Durand, Olivier — IFSTTAR
Tournat, Vincent — CNRS, Univ. du Maine, LAUM

Concrete is a widely used construction material by virtue of its cost and mechanical properties. Due to its low tensile strength however, concrete is very sensitive to crack formation. Cracks in concrete are responsible for significant inspection, maintenance and repair costs. In order to optimize structural health management, Non-Destructive Testing (NDT) has been extensively studied. Among all NDT techniques, ultrasonic methods are considered advantageous by providing information on mechanical properties in areas not directly accessible from the surface. Recent studies have led to developing nonlinear ultrasonic methods to increase the sensitivity to damage making possible the detection of large cracks/notches and the monitoring of crack evolution. However, the detection of small cracks in concrete remains a great challenge for NDT techniques.

In this study, an ultrasonic method, based on nonlinear acoustic mixing of coda waves by lower-frequency swept pump waves, providing for an efficient global detection of small cracks in concrete is presented. By simultaneous comparison, for uncracked and cracked mortars, of ultrasonic velocity variations and decorrelation coefficient between the unperturbed and the perturbed signals for different pump amplitude, the method allows to accurately detect very small cracks with widths of around 20 mm correlated with velocity variations of approximately 0.01%. This method is reproducible and able to provide a simple means for differentiating damaged and sound concrete. Attention must be paid however to the material evolution during the time span of both a single experiment and the entire experimental campaign as a consequence of the presumed high sensitivity of the observables.

Several applications of this technique could be developed in the field of civil engineering, although the power of the pump source would constitute a limitation. For example, the detection of small cracks causing leakage could be performed without any need for percolating fluid.

15:20-15:40 WeBT6.5

Monitoring and Imaging Based on Interferometric Concepts (I)

Sens-Schönfelder, Christoph	GFZ Potsdam
Niederleithinger, Ernst	BAM
Gassenmeier, Martina	GFZ Potsdam
Katja, Thiemann	K-UTEC
Köllner, Florian	GGL GmbH

Utilization of coherent phase information in complex wave fields formes the basis of interferometric time series analysis. The concept is known since decades, but until about 15 years ago there have been no practical implementations. Meanwhile seismic interferometry is used in a wide range from investigations of the earth's deep crust to engineering applications. Focused on monitoring and imaging the MIIC project as part of the German GeoTechnologien program has contributed to this development. Special attention was given to the transfer of methodology to different length scales that range from centimeters, in laboratory applications, over geotechnical scales to even kilometers in seismological applications. General purpose methods and open source software was developed, which can be used on all scales. The core of the MIIC software is a Python library organized in different modules for various processing tasks. A graphical user interface facilitates the creation of processing routines by visualizing connections and dependencies of variables and by checking the consistency of data types. Example applications have included carbon sequestration, salt mine and railroad embankment monitoring as well as imaging changes in concrete constructions.

15:40-16:00 WeBT6.6

Probing Macro Cracks in Concrete *Using Diffuse Ultrasound (I) (oral Only)*

Payan, Cedric	Aix Marseille Univ.
Garnier, Vincent	LMA CNRS
Chaix, Jean-François	Aix Marseille Univ. LMA CNRS UPR 7051
Quiviger, Audrey	Aix Marseille Univ. LMA CNRS UPR 7051

Several difficulties are encountered in the use of acoustic techniques for the characterization of concrete in an industrial context. ISO and ASTM standards state that ultrasound frequencies in the range between 20 kHz and 150 kHz should be used. When the frequency is increased (up to about 600kHz), ultrasound measurements become more complex, in particular due to multiple scattering, resulting from the presence of aggregates similar in size and spacing to the ultrasound wavelength. However, one can show that the complex propagation of multiple scattered waves in concrete can be simplified into a standard diffusion law. The fit of the coda waves energy envelope with the solution of this equation, allows to estimate the diffusion parameters: the diffusivity, the dissipation and the arrival time of the maximum energy.The present study reports recent experimental/simulation results allowing to probe localized surface breaking cracks in concrete.

WeCT1 Auditorium

Fiber Optics II (Regular Session)

Chair: Habel, Wolfgang R.	BAM Berlin
Co-Chair: Frövel, Malte	INTA, Inst. Nacional de Técnica Aeroespacial

16:20-16:40 WeCT1.1

Influences of Bonding Materials on the Accuracy of Fibre Bragg Grating Strain Measurements

Zhang, Wei	Chongqing Univ.
Chen, Weimin	Chongqing Univ.
Shu, Yuejie	Chongqing Univ.
Lei, Xiaohua	Chongqing Univ.

To achieve accurate strain measurements using fibre Bragg gratings (FBGs), both the sensing and wavelength demodulation accuracy are worth considering. Due to the indispensability, influences of bonding materials on the measurement accuracy need to be clarified. To analyze the sensing accuracy of the FBG strain sensor, a simplified longitudinal strain transmission model is established to discuss the transfer coefficient issue. Results indicate that hard bonding materials can enhance the sensing accuracy by high strain sensitivity. But on the other hand, asymmetric transversal stress model reveals that hard bonding materials can lead to the distortion of the spectrum in a large load condition. Then, effects of distorted spectra on the wavelength demodulation accuracy are studied by the simulation. The simulation shows that distorted spectra can reduce the wavelength demodulation accuracy markedly. It can be concluded that to improve the accuracy of strain measurements, hard bonding materials are suitable for small strain measurements, and relatively soft bonding materials are better options for large strain measurements.

16:40-17:00 WeCT1.2

A New Embedded Full Strain Components *Sensor*

Lecieux, Yann	GeM, Univ. of Nantes
Lupi, Cyril	GeM, Univ. of Nantes
Leduc, Dominique	GeM, Univ. of Nantes
François, Marc	Faculté des Sciences de Nantes, Lab. GeM
Roche, Michel	GeM, Univ. of Nantes

A new concept of strain (or stress) tensor sensor is presented. Fully embedded in the structure, it measures the full 3D strain tensor while suppressing its own influence, i.e allowing to find the strain (or stress) that would exist without the sensor, thanks to the Eshelby theorem. Furthermore, this theorem proves that the strain field is homogenous in the spherical (or ellipsoidal) body of the sensor, leading to perfect conditions of use for the six Bragg grating glass fibers which measure the strain of the body. After a brief overview of the theory, both mechanical and optical, we will present the results of the test performed using a prototype. This device is a simplified version of the sensor (3 channels) submitted to a hydrostatic pressure in order to assess the precision and the repeatability of the measure.

17:00-17:20 WeCT1.3

Radiation Tolerant Fiber Optic Humidity Sensors for for High-

Energy Physics *Applications*

Berruti, Gaia Maria	Univ. of Sannio
Consales, Marco	Univ. of Sannio
Borriello, Anna	Inst. for Composite and Biomedical Materials, CNR
Giordano, Michele	Inst. for Composite and Biomedical Materials, CNR
Buontempo, Salvatore	Istituto Nazionale di Fisica Nucleare - Sezione di Napoli
Breglio, Giovanni	Univ. of Naples Federico II
Makovec, Alajos	Univ. of Debrecen
Petagna, Paolo	CERN
Cusano, Andrea	Univ. of Sannio

In this contribution we present investigations developed in the last years by our multidisciplinary research group concerning the possibility to use fiber grating based sensors for relative humidity monitoring in the Compact Muon Solenoid (CMS) experiment at CERN, in Geneva. In particular our research, firstly focused on the development of relative humidity fiber Bragg grating (FBG) sensors coated with micrometer thin polyimide overlays, has been recently extended to long period grating (LPG) sensors, coated with a finely tuned titanium dioxide ($TiO2$) thin layer (~100 nm thick). Experimental tests in the range [0-75] %RH and at different temperatures were carried out to assess the FBG and LPG humidity sensors performances in real operative conditions required in experiments running at CERN. Progressive irradiation campaigns with γ- ionizing radiations were also performed. Obtained results demonstrate the strong potentialities of the two proposed technologies in light of their future exploitation as robust and valid alternative to currently used commercial hygrometers in High Energy Physics (HEP) applications.

17:20-17:40 WeCT1.4

Smaller Diameter Optical Fibre Sensor for Automated Embedding in Composite Laminates

Voet, Eli	Ghent Univ. Department of materials science and engineerin
Ahmed, Tahira	Airborne Tech. Centre
Lindner, Eric	FBGS Tech. GmbH
Luyckx, Geert	Univ. Gent - Department of Material Science and Engineeri
Brodsjo, Anders	Airborne Tech. Centre
Degrieck, Joris	Univ. Gent

In the framework of the FP7-project SMARTFIBER (http://www.smartfiber-fp7.eu/) an optical fibre prototype has been developed with a cladding diameter of approximately 60 μm. Draw Tower fibre Bragg gratings (DTG®s) are inscribed in this fibre which show identical strain optic and thermo optic characteristics compared to the standard commercial available 125 μm version, moreover they show superb ultimate failure strain of more than 7%. This characteristic plays an important role when automatic fibre embedding processes are envisaged. In this paper, the developed 60 μm fibre prototypes are successfully been applied during automated fibre placement on prepreg material. The results of the calibration tests and of the strain evolution of the DTG® during automated fibre placement will be presented in this paper

17:40-18:00 WeCT1.5

Strain Monitoring and Detection *of Barely Visible Damage in Foam-Core Sandwich Structures*

Siivola, Juho Tapani	The Univ. of Tokyo
Minakuchi, Shu	The Univ. of Tokyo
Mizutani, Tadahito	Japan Aerospace Exploration Agency
Takeda, Nobuo	The Univ. of Tokyo

To detect indentation or low-velocity impact induced damage in foam-core sandwich structures, a fiber-optic distributed strain monitoring system is applied to the structures. An optical fiber is embedded into the sandwich structures and the strains are measured using a Rayleigh scattering based monitoring system which offers high resolution. Indentation loading tests with sandwich beam and panel structures are conducted to verify the monitoring ability of the system and compared with numerical analysis predictions. The monitoring system could measure the strain distribution with very high spatial resolution and allowed for observation of the strain formation throughout the whole indentation event. The detectability of the damage in panel structures with regards to the location and density of the optical fiber sensor network is also discussed. The used system could accurately detect slightest changes in residual strains indicating barely visible damage and thus allowed for good estimation of the location and size of damage in the core of the sandwich structures.

WeCT2 GH Room

Life Cycle Analysis (Regular Session)

Chair: Schoefs, Franck	Univ. of Nantes
Co-Chair: Orcesi, André	IFSTTAR

16:20-16:40 WeCT2.1

Non Destructive *Testing of Concrete: Transfer from Laboratory to On-Site Measurement*

Garnier, Vincent	LMA CNRS
Martini, Dominique	LMA-LCND Aix-en-Provence
Salin, Jean	EDF R&D
Fardeau, Vincent	SETRA
Sbartai, Mehdi	I2M
Breysse, Denys	I2M
Piwakowski, Bogdan	Ec. Centrale de Lille
Villain, Géraldine	IFSTTAR
Abraham, Odile	IFSTTAR
Balayssac, Jean-Paul	LMDC Toulouse

The evaluation of mechanical and chemical properties of concrete by non-destructive evaluation requires the adaptation of the techniques and of theirs exploitations. Under the ACDC national project, we developed a methodology for the characterization of structures in which the reality of the material and the environmental conditions are taken into account. Correlations laws from the laboratory between non-destructive measurements and characteristics of the concrete are readjusted by several procedures. We invert the non-destructive evaluation and exploit the complementarity of the four techniques implemented (ultrasonic, electromagnetic, resistive and capacitive) through a process of data fusion. We develop the steps of the pre auscultation and the auscultation and define an estimator with a quality intrinsic of data fusion. Inversion procedures were defined and tested on two industrial structures. They rely on the consistency of non destructive measures or on data obtained from destructive testing. We present methods and results, the selection criteria and the approach adopted for operation. Ours propositions are written in a recomandations book to guide the auscultation.

16:40-17:00 WeCT2.2

Comparison of Durability Indicators Obtained by Non Destructive Testing Methods to Monitor *the Durability of Concrete Structures*

Villain, Géraldine	IFSTTAR
Balayssac, Jean-Paul	LMDC Toulouse
Garnier, Vincent	LMA CNRS
Piwakowski, Bogdan	Ec. Centrale de Lille
Salin, Jean	EDF R&D
Fardeau, Vincent	SETRA
Dérobert, Xavier	IFSTTAR
Coffec, Odile	IFSTTAR
Joubert, Anaëlle	IFSTTAR

This paper deals with the use of non destructive testing methods (NDT) to assess indicators of concrete durability and mechanical properties of reinforced concrete structures. On site, NDT methods based on electromagnetic or ultrasonic wave propagation (such as radar, impact echo, ultrasonic transmission device...) are used because they are more or less sensitive to water content and mechanical properties depending on the method. It has been shown, in a former project [1, 2], that the NDT results called "observables" are linked to mechanical and durability indicators (Young's modulus, compressive strength, porosity and saturation degree). Meanwhile, the relationship between observables and indicators depends on the concrete mix design. A calibration protocol is then proposed to get this relationship for the right mix of the reinforced structure studied by using a minimal number of cores. The cores are non-destructively characterised in laboratory or used to determined reference indicators by standardised destructive methods. The aims of this paper are first to present the ND calibration protocol on cores and then to validate this proposed calibration protocol. To achieve this goal, some NDT results obtained on site and on the corresponding core are compared and durability indicators deduced from NDT calibration are compared with reference durability indicators.

17:00-17:20 WeCT2.3

Ply Level Uncertainty Effects *on Failure of Composite Structures*

Mukherjee, Shuvajit	Indian Inst. of Science
Ganguli, Ranjan	Indian Inst. of Science
Gopalakrishnan, S	Indian Inst. of Science
Cot, Léa Dominique	Inst. Clément Ader
Bes, Christian	Inst. Clément Ader

Composite materials are popular in aerospace engineering because of their light weight, high strength to weight ratio, stiffness to weight ratio and tailoring properties. However, the modeling of composites is still a challenging task because of its highly uncertain material properties which arise from the manufacturing process.The effect of material uncertainties in failure strength and reliability analysis for single ply and cross ply laminated composite subjected to only axial loading is studied in this work. Researchers have categorized the uncertainty at different scales. Meso-scale uncertainty, that is ply level uncertainty, is being considered here. In this paper, the ply level uncertainties are taken as basic random variables and the strength parameters of the composite are derived through uncertainty propagation considering both Tsai-Wu and Maximum stress criteria. Monte Carlo simulation is performed to quantify uncertainty effects.

17:20-17:40 WeCT2.4

*Establishing a Wear-*Related *Deterioration Model Based on Experimental Data*

Beganovic, Nejra	Univ. of Duisburg-Essen
Rothe, Sandra	Univ. of Duisburg-Essen
Söffker, Dirk	Univ. of Duisburg-Essen

System modeling with respect to reliability function is of increasing importance, since knowledge about reliability function enables its integration in control affecting the control strategies. Suitable operating strategies will lead not only to extended lifetime, but also give a possibility for predictive maintenance. The assessment of reliability function relies often on probabilistic theory. From probabilistic point of view it consequently involves mathematical assumptions, but physically mainly relies on knowledge about system's state-of-deterioration. This implies that statements about the state-of-deterioration, about predictive maintenance, as well as the ability to control the system's extended lifetime assume knowledge about the relation between system's load (applied to the system), system's stress (appearing within the system), and the resulting behavior of systems to the loads applied. This contribution establishes a parametric model from the experimental data generated from wear experiments, and compares the results to those results obtained for individual system generating data.

17:40-18:00 WeCT2.5

Design and Implementation of an Integrated Operations and Preservation Performance Monitoring System for Asset Management of Major Bridges

Cohen, Ben	Drexel Univ.
Ye, Shi	Drexel Univ.
Karaman, S. Gokhan	Drexel Univ.
Khan, Fuad	Drexel Univ.
Bartoli, Ivan	Drexel Univ.
Pradhan, Anu	Drexel Univ.
Ellenberg, Andrew	Drexel Univ.
Moon, Franklin	Drexel Univ.
Gurian, Patrick	Drexel Univ.
Antonios, Kontsos	Drexel Univ.
Minaie, Ehsan	Intelligent Infrastructure Systems
Young, Charles	Intelligent Infrastructure Systems
Lowdermilk, David	Intelligent Infrastructure Systems
Aktan, Emin	Drexel Univ.

A cross-disciplinary group of researchers at Drexel University, spanning several generations, disciplines and departments, has taken a holistic approach to infrastructure research and education. Specifically, the Drexel research team is exploring the development of a comprehensive maintenance and management plan for movable bridges. This includes the transformation of state of the art inspection protocols that leverage expert knowledge, past events and existing standards, manuals and recommendations. Drexel researchers are also investigating Non-Destructive Evaluation (NDE) strategies and their potential implementation on robotic platforms for automated assessment and evaluation of infrastructures to augment the inspection process. Thanks to the foresight of a regional infrastructure owner in the Greater Philadelphia Area, the described initiatives are allowing the rigorous implementation of technology on real, complex systems. Furthermore, many of these research efforts will be integrated into existing courses at Drexel University where students will be exposed to real life living infrastructures.

WeCT3 Room I

Realtime Methods (Regular Session)

Chair: Uhl, Tadeusz	AGH Univ. of Science and Technology, Kraków
Co-Chair: Le Cam, Vincent	IFSTTAR

16:20-16:40 WeCT3.1

Quality Assessment of Dynamic Response Measurements Using Wireless Sensor Networks*: Preliminary Results*

Chowdhury, Sharmistha	Bauhaus-Univ. Weimar
Olney, Peter	Bauhaus-Univ. Weimar
Deeb, Maher	Bauhaus-Univ. Weimar
Zabel, Volkmar	Bauhaus-Univ. Weimar
Smarsly, Kay	Bauhaus-Univ. Weimar

Vibration-based methods, using response measurements of civil engineering structures, are among the most popular methods for damage detection and system identification in structural health monitoring (SHM). To accurately obtain response measurements, wireless sensors networks are increasingly applied for collecting monitoring data from the observed structures. However, several types of noise usually contaminate the monitoring data leading to inaccurate or even incorrect diagnoses and prognoses of the dynamic properties of the structures. The main objective of this study is the elimination of such noise directly on the wireless sensor nodes in order to obtain high-quality monitoring data. Specifically, a wireless SHM system is designed and, for validation purposes, mounted on a tower-like laboratory test structure. The wireless sensors are provided with different embedded filters designed to improve the estimation of the system properties, which are calculated from the dynamic response measurements, i.e. from the monitoring data. In a final step, the quality of the dynamic response measurement is assessed.

16:40-17:00 WeCT3.2

Real-Time Active Pipeline Integrity Detection (RAPID) System for Corrosion Detection and Quantification

Bergman, Jeffrey D.	Acellent Tech.
Lee, Sang Jun	Acellent Tech.
Chung, Howard	Acellent Tech.
Li, Irene	Acellent Tech.

Structural Health Monitoring (SHM) technologies offer a paradigm shift from schedule driven inspection and maintenance to on-demand inspection and Condition Based Maintenance (CBM). Utilizing SMART Layer technology and lamb-wave based damage detection Acellent has developed a Real-time Active Pipeline Integrity Detection (RAPID) system. The RAPID system utilizes a sensor network permanently bonded to the pipeline structure along with in-situ networked hardware and remote access and damage detection programs to provide both scheduled and on-demand monitoring of pipeline structures. Advantages of the RAPID system include: 1) Automated and on-demand inspection of critical areas; 2) Damage localization and quantification; 3) Easy to use interface requiring minimal training. To verify the capabilities of the system a series of tests were performed by Acellent in partnership with Chevron utilizing sections of 8in diameter steel pipes. During the tests a number of different sizes and depths of defects were introduced into the pipeline sections. These tests verified that the RAPID system was effective in detecting the occurrence of corrosion in the pipeline and monitoring its growth over time.

17:00-17:20 WeCT3.3

Sequential Importance Sampling Based on a Committee of Artificial Neural Networks for Posterior *Health Condition Estimation*

Sbarufatti, Claudio — Pol. di Milano

The output of real-time diagnostic systems based on the interpretation of signals from a sensor network is often affected by very large uncertainties if compared with local non-destructive testing methods. Sequential Importance Resampling (SIR) is used in this study to filter the output distribution from a committee of Artificial Neural Networks. The methodology is applied to a helicopter panel subject to fatigue crack propagation. Strain signals are acquired during crack evolution and a diagnostic unit trained on simulated experience provides damage assessment in real-time. This information is filtered through a SIR routine, providing model identification, model parameter estimation and crack length probability density function updating, conditioned on the observations at discrete time steps.

17:20-17:40 WeCT3.4

Structural Damage Detection Algorithm Based on Principal Component Indexes and Embedded *on a Real Time Platform*

Villamizar, Rodolfo — Univ. Industrial de Santander
Camacho, Jhonatan — Univ. industrial de Santander
Mujica, Luis Eduardo — Tech. Univ. of Catalonia
Ruiz, Magda — Univ. of Girona (UdG),
Quiroga, John Leonardo — Univ. Industrial de Santander

This paper presents the main results obtained by using a structural damage detection algorithm based on Principal Component Analysis (PCA) and piezo-actuation principle. A known high frequency piezoactuated signal is applied on an analyzed structure in order to determine the base-line performance for the undamaged state (undamaged PCA model). Q-statistic and hotteling's T2 indexes are computed by projecting time data onto the principal component space, and used to identify deviations of the current dynamical responses respect to the undamaged state. The algorithm was embedded in the Beaglebone Black Hardware (platform based on an ARM cortex A8 processor) and tested by using experimental data supplied by the CODALAB group. The obtained results indicate that it is possible to identify and locate structural faults for this kind of structures. Identification capability of the algorithm for 10 damages is tested by adding masses on the surface of an aircraft turbine blade at different positions.

17:40-18:00 WeCT3.5

Embedded Multi-Tone Ultrasonic Excitation and Continuous-Scanning Laser Doppler *Vibrometry for Rapid and Remote Imaging of Structural Defects*

Flynn, Eric — Los Alamos National Lab.

We describe a novel method for rapidly measuring local wave dispersion properties using steady-state excitation continuous-scanning laser Doppler vibrometery (CSLDV). In our approach, we excite a structure with a periodic ultrasonic waveform constructed from the sum of several single-tone waveforms. The structure is excited continuously, bringing it to steady-state. We then measure the steady-state response of the structure through CLSDV. The continuous scan gives a one-dimensional time-history of measured response velocity, which we convert to time histories of the instantaneous amplitude and phase for each excitation frequency component. We then map these instantaneous amplitudes and phases to a discretized grid of points spanning the scan area. Finally, through wavenumber processing, we convert these 2D maps into local estimates of wavenumber. Since this is done for each excitation frequency, we are left with a set of frequency samples of local frequency-wavenumber dispersion curves at each spatial sample point in the structure. Since defects alter local dispersion properties in a predictable way, these dispersion curve maps offer an effective means of both locating and characterizing defects such as corrosion and delamination.

WeCT4 — Room J

Pattern Recognition III (Regular Session)

Chair: Omenzetter, Piotr — Univ. of Aberdeen
Co-Chair: Mevel, Laurent — Inria

16:20-16:40 WeCT4.1

A Structural Damage Detection Indicator Based on Principal Component Analysis *and Multivariate Hypothesis Testing Over Scores*

Pozo, Francesc — Univ. Pol. de Catalunya
Arruga, Ignacio — Univ. Pol. de Catalunya-BarcelonaTech
Mujica, Luis Eduardo — Tech. Univ. of Catalonia

This paper is focused on the development of a damage detection indicator that combines a data driven baseline model (reference pattern obtained from the healthy structure) based on principal component analysis (PCA) and multivariate hypothesis testing. More precisely, a test for the plausibility of a value for a normal population mean vector is performed. The results indicate that the test is able to accurately clasify random samples as healthy or not.

16:40-17:00 WeCT4.2

Benchmark Data for *Structural Health Monitoring*

Kullaa, Jyrki — Helsinki Metropolia Univ. of Applied Sciences

Data analysis is a key function in structural health monitoring (SHM). To develop algorithms for SHM, one needs realistic data. A library of SHM data is introduced with simulations of vibration measurements resulting in three challenging SHM cases: (1) a beam structure with environmental and operational influences, (2) a beam with a non-linear breathing crack, and (3) moving loads on a beam, modelling traffic on a bridge. In all these cases, the excitation is varying and unknown. Also, the environmental or operational variables are random and unknown. A different modelling strategy is used in each case. Also the damage scenarios are different. Data analysis examples for damage detection and localization are presented utilizing redundancy of the sensor network. There are endless possibilities to utilize the benchmark data.

17:00-17:20 WeCT4.3

Wavelet Based Characterization of Acoustic Attenuation in Polymers Using Lamb *Wave Modes*

Ahmad, Rais — California State Univ. Northridge

Polymers have been used in a wide range of applications ranging from fabrication of sophisticated medical equipment to manufacturing aircrafts. The design advantages of using polymers are its high strength-to-weight ratio, resilience, and compatibility with net-shape processes. In recent years, researchers have been trying to ascertain the mechanical as well as acoustical properties of polymers. Acoustical properties like attenuation of propagating ultrasonic waves through polymers vary in a broad spectrum depending on their chemical structure and stoichiometry. Guided wave techniques are widely used for nondestructive evaluation and inspection as well as examining the integrity of structures.

This study demonstrates that guided wave techniques can be effectively utilized for material characterization, where efficient characterization depends on identification and selection of appropriate propagating wave modes and suitable signal processing techniques. The focus of this investigation is to estimate acoustic attenuation of acrylic (PMMA, polymethyl methacrylate), thermoplastic, using guided Lamb wave. Lamb waves are generated and received by piezo-electric transducers in a pitch-catch configuration. The received signals are first isolated from the inherent white noise using db4 based wavelet algorithm. The de-noised signals are then processed using Gabor Transform, which provides information about the group velocities of the propagating modes. The experimentally determined group velocities are compared with theoretical group velocities of the investigated polymers to identify the propagating Lamb wave modes. An effort has been made to estimate the attenuative properties of the thermoplastic from selective propagating Lamb wave modes.

17:20-17:40 WeCT4.4

An Unsupervised Pattern Recognition Approach for AE Data Originating from Fatigue Tests on Polymer-Composite Materials

Doan, Dinh Dong — Femto-ST Inst.
Ramasso, Emmanuel — Femto-st Inst.
Placet, Vincent — Femto-st Inst.
Boubakar, Mohamed Lamine — Femto-st Inst.
Zerhouni, Noureddine — FEMTO-ST Inst. UMR CNRS 6174 - UFC / ENSMM / UTBM

Acoustic Emission (AE) technique is gaining more and more interest for structural health monitoring (SHM) in polymer-composite materials. Recent literature has shown that using appropriate pattern recognition techniques (PRT), the identification of the natural clusters of acoustic emission data can be obtained. This work investigates acoustic emission generated during tension fatigue tests carried out on a carbon fiber reinforced polymer (CFRP) composite specimen. Since fatigue data processing, especially noise reduction remains an important challenge in AE data analysis, a noise modeling has been proposed in the present work to tackle this problem. A Davies-Bouldin-index-based progressive feature selection has been implemented to reduce high dimensional fatigue dataset. A classifier offline-learned from quasi-static data is then used to classify the processed data to different AE sources. An adaptation has been studied to enable the classifier to generate new class, i.e. AE source, for unidentified AE events. With efficient proposed noise removal and automatic separation of AE events, the results of this work provide an insight into fatigue damage development in composites and then ability to health assessment which is necessary for residual life prediction. KEYWORDS: organic-matrix composites, acoustic emission, data clustering, noise reduction, feature selection.

17:40-18:00 WeCT4.5

Finite Wavelet Domain Method for Efficient Modeling of Lamb Wave Based Structural Health Monitoring Systems

Theodosiou, Theodosis — Univ. of Patras
Nastos, Christos — Univ. of Patras
Rekatsinas, Christoforos — Univ. of Patras
Saravanos, Dimitris — Univ. of Patras

This document describes the development of an innovative and computationally efficient modeling approach for the prediction of the transient response in elastic rods and strips. The introduced beam element exploits the advantages of wavelets for the spatial discretization of the displacement field. Results are validated against with confirmed models and are found to be in agreement. The model appears to be very accurate and computationally more efficient than other popular numerical methods.

WeCT5 KL Room

Modal Analysis (Regular Session)

Chair: Tcherniak, Dmitri — Bruel and Kjaer Sound and Vibration Measurement

Co-Chair: Döhler, Michael — Inria

16:20-16:40 WeCT5.1

Optimal Sensor Placement Methodology for Operational Modal System Identification of a Hyperbolic Paraboloidal Fabric Structure

Castro Triguero, Rafael — Univ. of Cordoba
Rodríguez León, Mª Teresa — Univ. of Seville
García Macías, Enrique — Univ. of Córdoba
Gallego Sevilla, Rafael — Univ. of Granada
Sánchez Sánchez, José — Univ. of Seville
Vázquez Vicente, Enrique — Univ. of Seville

This paper presents a numerical pre-test that considers finite element modelling and optimal sensor placement study for a singular spatial truss structure field vibration test. The singularity and importance of such structure require easier, quicker and cheaper monitoring methods. Vibration-based health monitoring methods determine the modal characteristics of the structure via a limited number of sensors. These characteristics are intrinsic properties, so that a variation in them may be induced by structural damage. Only a limited number of degrees-of-freedom can be measured for the system identification process. By developing a finite element model for the spatial truss structure, these degrees-of-freedom can be identified. Based on these results, Effective Independence Method (EFI) method is employed to determine the optimal sensor number and locations. This methodology is formulated with the use of the modal properties of the structure, where the independence of the target modal shape matrix is maximised in an iterative process, and those degrees-of-freedom that do not contribute to the independence of the target modes are eliminated. It is concluded that utilising just a few number of sensors the result of placement location is improved for structural health monitoring purposes.

16:40-17:00 WeCT5.2

Vibration-Based Monitoring of Tensile Loads: System Development and Application

Rainieri, Carlo — Univ. of Molise
Gargaro, Danilo — Univ. of Molise
Fabbrocino, Giovanni — Univ. of Molise

The in-situ evaluation of the tensile load in cables and tie-rods plays a primary role in the continuous monitoring and health assessment of strategic as well as historic structures. The main advantage of the indirect determination of cable forces from dynamic tests in operational conditions is the opportunity to achieve accurate, cheap and fast quality checks in the construction phase (after pre-stressing), and safety checks and structural maintenance over the structure lifespan. Based on the most recent developments in the field of operational modal analysis, an automated system for continuous monitoring of axial loads based on dynamic measurements has been developed. The system consists of a distributed measurement system based on programmable hardware and wireless modules, and of a centralized data processing server for the estimation of the tensile loads in the monitored cable from records of its dynamic response to ambient vibrations. The experimental identification of the flexural modes in terms of natural frequencies and mode shapes is used to solve an inverse problem for the identification of the axial loads. A prototype of the SHM system has been installed to monitor the tensile load in one of the cables of a sample steel arch. Preliminary experimental results from continuous monitoring of the cable are discussed pointing out the effect of environmental factors.

17:00-17:20 WeCT5.3

SHM Based on Modal Analysis: Accelerometer and Piezoelectric Transducers Instrumentation for Civil Engineering in Heterogeneous Structures

Andrés, Belisario — Lab. d'Analyse et d'Architecture des Systèmes

This paper presents a strategy for the detection of mechanical damage of a reinforced concrete beam strengthened by Carbon Fiber Reinforced Polymers (CFRP) sheets (TFC ©) based on the joint use of piezoelectric sensors and conventional accelerometers. The beam is subjected to repeat impacts that may represent the action of vehicles on a roadway joint of a bridge, at different levels of damage induce by bending of the reinforced concrete beam. The results show that piezoelectric sensors provide a much wide response those accelerometers allowing a study in the high frequency range (above 2 kHz). The spectral response appears largely affected by mechanical damage on the beam and suggests a possible use for SHM in Civil Engineering structures.

17:20-17:40 WeCT5.4

Modal Analysis of Reinforced Concrete Buildings Using Multipath Lidar

Gueguen, Philippe ISTerre/UGA/CNRS/IFSTTAR
Valla, Matthieu ONERA
Augere, Béatrice ONERA
Goular, Didier ONERA

This paper compares the modal analysis of reinforced-concrete buildings obtained using sensitive velocimeters and coherent LIDAR. Ambient vibrations are recorded by these two systems and processing using operative modal analysis method for getting building frequency and mode shapes. Real-scale trials applied to five buildings located at Grenoble (France) are presented. The efficiency and reliability of the Lidar is discussed and the modal parameters measured by Lidar at a range of 200m and by in-situ velocimeters are compared. The results are in good agreement and allow us to conclude on the ability of the coherent Lidar to assess modal parameters of existing buildings at long range and without any retroreflectors placed on the structures. The results open new perspectives for remotely testing buildings, without getting inside, facilitating dynamic analysis of buildings for earthquake engineering applications.

17:40-18:00 WeCT5.5

Vibration Amplitude - Dependent *Natural Frequency and Damping Ratio of Repaired Pier Model*

AlSehnawi, Reem Utsunomiya Univ.
Nakajima, Akinori Utsunomiya Univ.
Takeshima, Ryuji IHI Infrastructure Systems Co., Ltd.
Al Sadeq, Hafez Damascus Univ.

The present work aims to investigate the dependency of dynamic characteristics such as the natural frequency and the damping ratio of pier model on the vibration amplitude under a series of different vibration levels. Several free vibration tests with a wide range of amplitudes were, therefore, carried out under excitation with relatively small amplitude levels. The pier model was subjected to a progressive damage by conducting a shaking table test to study the influence of different damage scenarios on its dynamic characteristics. After the damage occurs at the plastic hinge zone of the pier model, free vibration tests with different amplitude excitation were carried out to derive the dynamic characteristics of the damaged pier model. Furthermore, after the damage of the pier model was repaired by carbon fiber sheets (CFS) jacketing, free vibration test was carried out again, to derive the dynamic characteristics of the repaired pier model. Generally, this study indicates that overall natural frequency and damping ratio are very sensitive to vibration amplitude level. Particularly, whenever the acceleration amplitude increases, damping ratio significantly increases while natural frequency slightly decreases. Also, the repairing method by CFS proved to be effective in restoring performance characteristics of damaged pier.

WeCT6 Room 200
Nonlinear Acoustics (Regular Session)

Chair: Saravanos, Dimitris Univ. of Patras
Co-Chair: Giurgiutiu, Victor Univ. of South Carolina

16:20-16:40 WeCT6.1

Experimental Investigation of Fatigue Damage in Composite Structures Considering *Second Harmonic Lamb Waves*

Rauter, Natalie Helmut-Schmidt-Univ. / Univ. of the Federal Armed Forc
Lammering, Rolf Helmut-Schmidt-Univ. der Bundeswehr Hamburg

To detect microstructural damages like fiber / matrix cracks and delaminations in composite materials accurately new methods are developed. Previous works have investigated and shown that the acoustical nonlinearity parameter is an appropriate tool to detect micro-structural damages like plasticity and fatigue in metal as well as thermal fatigue and impact damages in composites. In this work the second harmonic generation is used to analyze unidirectional composite specimens for fatigue damage caused by cyclic tensile load. Therefore, Lamb waves are launched and detected by piezoelectric actuator and sensor, respectively, at a certain frequency to generate cumulative second harmonic modes. The excitation frequency has to meet special conditions. The signal processing is done by using the wavelet transform. The correlation of the results of the nonlinear wave propagation to the damage state is investigated. It is shown that the relative acoustical nonlinearity parameter is sensitive to the damage state.

16:40-17:00 WeCT6.2

Lamb Wave Propagation in Complex Geometries - Model Reduction with Approximated *Stiffeners*

Heinze, Christoph German Aerospace Center (DLR)
Sinapius, Michael Tech. Univ. Braunschweig, Inst. of Adaptronics an
Wierach, Peter German Aerospace Center (DLR), Inst. of Composite Structures

Aviation companies show great interest in Structural Health Monitoring (SHM) systems. Although noticeable effort has been put into this research field, so far no system is ready for industrial use. One of the major issues is the processing of complex signals caused by complex aerospace structures. Numerical simulations can help to comprehend these signals, but are often unable to cope with the model detail required. A ray tracing algorithm to approximate Lamb wave propagation is proposed, which enables a huge model reduction. The resulting minimal model mainly consists of the border lines of areas with different geometrical or material properties. Finite element simulations are used to determine the interaction behavior of these transition zones. A two stage concept is used to find paths from an actuator to a sensor and calculate time signals.

17:00-17:20 WeCT6.3

Nonlinear Ultrasonic *Waves for Structural Health Monitoring*

Nucera, Claudio Univ. of California San Diego
Lanza di Scalea, Francesco Univ. of California San Diego

Nonlinear ultrasonic waves in solids have shown good sensitivity to quasi-static stresses. The stress sensitivity of elastic waves is typically associated to finite strains (e.g. theory of acoustoelasticity). In the case of waveguides, classical nonlinear theories for guided waves are still based on the assumption of finite strains. In the case of constrained solids subjected to thermal excursions, however, there are theoretically no finite strains (for perfectly-constrained solids) associated with thermal stresses. A new model is therefore needed to justify the existence of wave nonlinearities in this case of stress without strain. This problem is solved on the basis of the interatomic potential of the solid that indicates a "residual" strain energy, due to the prevented thermal expansion, that is at least cubic as a function of strain. The cubic relationship between strain energy and strain produces second-harmonic generation of propagating elastic waves. Consequently, a nonlinear wave equation can be derived. The solution to this equation leads to a new nonlinear parameter for double harmonic generation that is directly related to the thermal stresses in the structure. The present study finds applications in the monitoring of thermal stresses in buckling-prone structures, such as continuously-welded railroad tracks and pipelines.

17:20-17:40 WeCT6.4

Nonlinear Acoustic Imaging *of Structural Damages in Laminated Composites*

Pieczonka, Lukasz AGH Univ. of Science and

Technology, Kraków

Klepka, Andrzej — AGH Univ. of Science and Technology, Kraków

Staszewski, Wieslaw J. — Department of Robotics and Mechatronics, AGH Univ. of Science and Technology, Kraków

Uhl, Tadeusz — AGH Univ. of Science and Technology, Kraków

The paper describes the application of nonlinear acoustic techniques recently reported in scientific literature for imaging of structural damage in a laminated composite plate. The techniques that have been considered include the local defect resonance technique and the second harmonic imaging technique. The tests are performed on a carbon fiber/epoxy laminated composite plate with barely visible impact damage that was generated in an impact test. The extent of damage is characterized by vibrothermography prior to nonlinear acoustic tests. The paper discusses theoretical background and experimental setup for each measurement technique.

17:40-18:00 WeCT6.5

Numerical Computation of the Nonlinear Far Field of Ultrasonic Waves Scattered *by Closed Cracks of Various Orientations*

Blanloeuil, Philippe — Univ. of Queensland

Meziane, Anissa — Univ. of Bordeaux, I2M, UMR 5295, F-33400 Talence, France

Norris, Andrew N. — Rutgers Univ.

Renier, Mathieu — Univ. Bordeaux, I2M, UMR 5295, F-33400 Talence, France.

Veidt, Martin — Univ. of Queensland

The directivity patterns of the higher harmonics generated by the interaction between a bulk wave and a closed crack are obtained in the far field. A Finite Element (FE) model gives the near field solution which is then analytically propagated with expressions based on Hankel functions. The crack is modeled by an interface of unilateral contact with Coulomb's friction which takes into account a compression pre-stress s0 that closes the crack. The mode conversion occurring at the crack investigated. Moreover, the diffracted fields of the generated higher harmonics are compared to those of the fundamental harmonic.

Thursday July 10, 2014

Thursday July 10, 2014

Keynote Lectures Auditorium

Chair: Schoefs, Franck — University of Nantes

08:30-09:00 ThKN1L

Track Condition Monitoring at SNCF : An Introduction to Methods, Tools and Perpectives

L'Henoret, Benjamin — SNCF Company
Gigleux, Hughes — SNCF Company

SNCF INFRA has a long history of condition monitoring of its assets. This keynote will describe the methods used to gather the information today. As digital revolution opens new perspectives to always improve this monitoring, this keynote will briefly introduce some of the evolutions developed by SNCF to embrace this challenge, including big data to monitor track geometry with commercial trains or new trains doing automatic image processing.

09:00-09:30 ThKN2L

Improvement of Civil Works *Health Management: EDF's Global Strategy and Major Advances between 2008 and 2013*

Stephan, Pierre — EDF Company
Salin, Jean — EDF Company

EDF owns different types of civil works : some are of classical design (office buildings for example) and others are of particular design (containment building, cooling towers, dams..) since they have to withstand high mechanical loads and sometimes to meet other technical specifications (limited leak flow rate for example) during dozen of years. This is important because that means that EDF has to face the general damages encountered in classical civil works (especially corrosion) and the particular issues of its specific structures.

As responsible civil works owner, EDF has obviously to guarantee the safety of its structures. To do so, EDF instruments its civil works and follows a methodology to monitor their structural health. This methodology is first based on the knowledge of each structure: all available information about them must be gathered (design, construction, operation, external events...). Then an inspection program (continuous and periodic monitoring) allows the establishing of a simulated and a real image of the structures through the collection of visible damages and the result of the re-calculation of the structures' mechanical resistance, following the re-assessment of the structures' materials properties. The comparison of these two images leads to a structural diagnosis and prognosis whose aim is to take a decision about the eventual repair of the concerned civil work. The knowledge of the structure is then updated.

As the monitoring of civil works and the understanding of reinforced concrete damages (corrosion for example) is nowadays not always mastered, the improvement of this methodology requires studies in each of its steps (numerical simulations and models, inspection methods and means, damages indicators, prognosis...). This is not cheap. Moreover, in parallel, EDF has to improve its knowledge about the particular issues of its specific structures.

So EDF R&D has put in place, since years, a strategy to address this, based on:

- An active involvement in French national projects, to advance in knowledge on common issues (material, models, damages, ND techniques, Risk Base Analysis) for all civil engineering works,
- An active involvement in international projects, in collaboration with other Power plants owners, in order to work on common issues encountered in their specific structures (reactor containments, cooling towers, dams...).

This presentation will detail this strategy and give the main results issued from the numerous projects EDF took part between 2008 and 2013.

09:30-10:00 ThKN3L

Structural Health Monitoring of Civil Engineering Structures - Lessons Learned *and Future Directions*

Ventura, Carlos — British Columbia University

In recent years, Structural Health Monitoring (SHM) of Civil Engineering Structures has attracted considerable attention. SHM for Civil Engineering applications can be defined as "the process of acquiring and analyzing data from on-site sensors to evaluate the health of a structure." The suite of on-site sensors could include any presently installed sensors and new sensors to be installed in the future. The development of SHM technologies for Civil Engineering Structures is an activity that spans multiple engineering disciplines. It is also recognized that many stakeholders, like owners of infrastructure, regulatory agencies, designers, contractors, academia, and equipment suppliers are crucial to the process of certifying viable SHM solutions. Significant advances have been made in the theory related to the "health" of structures, and the present engineering applications of SHM are very encouraging and promising. However, a large number of methods and techniques have been proposed and applied to different structures, with different levels of success. This large variety of approaches makes it difficult to compare and contrast the merits of the various methodologies. And at the present time it is not possible to determine with certainty which methods and techniques are the most suitable for monitoring large structures like bridges, buildings and dams. This has also caused significant level of confusion to the owners and operators of infrastructure as it is not clear to them what SHM entails and what should be their realistic expectations when a structural health monitoring program is implemented. As a consequence the use of SHM for large structures has not become as prevalent as it had been expected when the concept was developed a few decades ago.

This presentation will provide a general overview of the various monitoring methodologies that are of practical applicability to Civil Engineering Structures and will provide provoking thoughts of what the academic and research community should do to encourage owners and operators of Civil Engineering structures to take full advantage of this technology to monitor and protect their assets. The need for guidance on the definition, development and certification of Structural Health Monitoring (SHM) technologies for Civil Engineering health management applications will be discussed as well.

ThAT1 Auditorium
Strain Monitoring I (Regular Session)

Chair: Casari, Pascal — Univ. of Nantes
Co-Chair: Alexandre, Paleologue — CEA Leti

10:30-10:50 ThAT1.1

Repair Patch Monitoring with Embedded Optical Sensors by the Residual Strain Release

Frövel, Malte — INTA, Inst. Nacional de Técnica Aeroespacial
Sanmillan, Javier — INTA
Maroto, Jose — INTA
Pintado, Jose Maria — INTA
Kressel, Iddo — Israel Aerospace Industries
Tur, Moshe — Tel-Aviv Univ.
Nachum, Gorbatov — Tel-Aviv Univ.

Repair of metal structures with carbon fibre reinforced plastic (CFRP) patches has a high potential for aeronautic structures. Bonded repairs with CFRP offer the advantage within others to repair in places where a standard riveted repair is not possible due to space limitations or geometrical conditions. It is of vital importance that a bonded repair patch will not peel off during service. Monitoring the patches provides a tool to detect a debond of the repair patch before the debond becomes a critical issue. A monitoring technique has been studied that enables to detect a patch peel off using fiber optic Bragg grating sensors (FBGS) embedded in the CFRP-patch. Temperature compensation has been introduced to enable the application of this technique in a wide temperature range. The technique allows the monitoring of the patch on ground after a defined number of flight cycles. The viability of the monitoring principle was studied and demonstrated on seven specimens made of typical aeronautic aluminum with bonded CFRP patches that are instrumented with two or four embedded FBGS. The results of the tests show that the embedded sensors can indeed sense the debonding of a bonded CFRP patch used to repair an aluminum structure. The sensitivity of the sensors is sufficiently high to detect a crack front that is still 10 mm far from the sensor. The temperature effect on the embedded sensors that monitors the crack front can be compensated by a second sensor that is far from the sensor front. A robust pre-cured sensor mat design was developed and manufactured. This sensing mate was made using four three fibers each with four FBGSs, embedded between two glass fabric plies, providing repeatable, easy and robust sensor integration. Test structures with monitored composite repair pat

10:50-11:10 ThAT1.2

The Health Monitoring of a Prestressed Concrete Beam Using Inverse Modeling Technique and Measured Dynamic Response

Smail, Mostafa Kamel — IFSTTAR
Waeytens, Julien — Univ. Paris-Est, IFSTTAR

This paper presents an inverse modeling technique using dynamic strain responses to identify the prestressed concrete beam parameters: the pretension in the cable and the prestress in the concrete. The direct problem corresponds to 2D elastodynamics equations. To get the beam parameters, we minimize a classical data misfit function using a gradient-like algorithm. A low-cost computation of the functional gradient is performed using the adjoint problem

11:10-11:30 ThAT1.3

Adaptable Synchronous Detection for the SHM of Composite Component: Autonomous FPAA *Architecture on Polyimide Film*

Zedek, Sabeha — LAAS

Structural Health Management strategies usually involve monitoring complex components, during their working life. Carbon fiber reinforced polymer components are becoming more and more used, throughout the structural engineering fields, because of their superior mechanical properties. Moreover, due to their construction, these materials open the path for new "smart" components, with integrated sensors and monitoring capabilities. Flexible circuits are essential in this context. In order to adapt to complex geometries, sensors and their connectivity have to be flexible and as un-intrusive as possible. Following previous work, aimed at inserting sensors with flexible connectivity in between the sheets of carbon fibers, we conceived and implemented a reconfigurable acquisition system that can be easily adapted to perform strain measurements in various EM5 noise conditions. This paper presents a comparison between three circuits dedicated to strain measurement using a commercial strain gauge and flexible connectivity on Kapton2. The first circuit consists of a classic Wheatstone bridge under continuous voltage. The second uses the synchronous detection principle with classic circuits and functions. The third innovates upon the synchronous detection principle by implementing several functions in a reprogrammable analog chip called FPAA (Filed Programmable Analog Array). The analysis of these circuits shows the added value of using reconfigurable devices with flexible connectivity and sensors in order to provide a universal solution for the monitoring of complex components.

11:30-11:50 ThAT1.4

Dynamic Strain Prediction *Using Modal Parameters*

Gevinski, Jakerson Ricardo — Univ. of Campinas
Pederiva, Robson — Univ. of Campinas

Vibration monitoring is conventionally performed from measurements of acceleration, velocity or displacement. However, these are not the primer parameters of interest in most structural diagnosis. Dynamic stress and strain are the appropriate parameters, which may be used in this monitoring. In some cases, strain is measured using strain gages techniques. But, when it is required to measure strain at different points, these techniques become onerous, because the strain gages must be fixed in structure and cannot be reused. In this sense, some methods have been developed to predict the dynamic strain from vibration measurements on structures. These methods basically consist in the numerical differentiation of displacement, obtained by modal analysis. Here, the dynamic strain prediction, based on hybrid modal analysis and acceleration measurements, is carried out. We used the finite difference method in the displacement to strain transformation. The displacements were obtained from measurements of acceleration and operating deflection shapes technique. The predicted strains were compared with the measured strains in the time domain. The predicted results closely agree with the measured results.

11:50-12:10 ThAT1.5

Strain Monitoring in Stiffened Composite Panels Using Embedded Fibre Optical and Strain Gauge Sensors

Ruzek, Roman — VZLÚ
Kadlec, Martin — VZLÚ
Tserpes, Konstantinos — Univ. of Patras
Karachalios, Evanggelos — Hellenic Aerospace Industry

This paper describes an implemented structural health monitoring (SHM) system based on Fibre Optic Bragg Grating (FOBG) sensors and standard resistance strain gauges (SGs) placed onto/into a stiffened carbon fibre-reinforced polymer (CFRP) fuselage panel. The role of the FOBG sensor system was to monitor the structural integrity of the reference, impacted and fatigued panels under compression loading. The FOBG and/or SG sensors were placed (embedded or bonded) at various locations into and/or onto the structure. The functionality of the SHM system was verified by experimental evaluation. The comparability and reliability of the results gained from the FOBG sensors and resistance strain gauges is discussed. This paper presents the key objectives from a European Union funded collaborative Project, inside the Clean Sky Joint Technology Initiative (CS JTI), investigating the SHM systems for CFRP fuselage stiffened panels based on fibre optic Bragg grating sensors for the Green Regional Aircraft Integrated Technology Demonstrator (GRA ITD).

ThAT2 — GH Room
Statistical Approaches I (Regular Session)

Chair: Glisic, Branko — Princeton Univ.
Co-Chair: Bernal, Dionisio — Northeastern Univ.

10:30-10:50 ThAT2.1

Reliable Identification *of Damage Growth Using Guided Wave SHM Systems*

Cawley, Peter — Imperial Coll.
Galvagni, Andrea — Imperial Coll.

A practical guided wave SHM system will be interrogated frequently and this will generate a large volume of data that must be processed. This paper discusses the use of sequential analysis to assess whether a change has occurred and whether it is of a form consistent with damage growth, rather than environmental changes. A methodology based on the generalised likelihood ratio (GLR) algorithm is introduced and applied to the guided wave monitoring of an 8 inch diameter pipe loop which is subjected to temperature cycling and the introduction of damage at different locations along the pipe. It is shown that the GLR algorithm can be used to process the signals automatically and to identify damage at specified probability of detection (POD) and probability of false alarm (PFA) rates. It is possible to detect damage corresponding to 0.25% cross section area loss in regions remote from features and to 0.5-0.75% loss at other locations. The methodology promises to be an important tool in the practical implementation of guided wave SHM.

10:50-11:10 ThAT2.2

Damage Detection and Localisation Using Mode-Based Method and Perturbation *Theory*

Hamze, Alaa	ISTerre/UGA/CNRS
Gueguen, Philippe	ISTerre/UGA/CNRS/IFSTTAR
Roux, Philippe	ISTerre/UGA/CNRS
Baillet, Laurent	ISTerre/UGA/CNRS

In this paper, the detection and the localization of a local perturbation are assessed by analysing the frequency changes only (fundamental mode and overtones). After describing the method used herein applied to the bending beam and based on the perturbation theory, experimental application to a 1D plexiglas beam is shown using frequency and modal analysis technique. The damage is considered as a local perturbation of Young's modulus. Finally, the localisation of damage is done using classical modal-based methods and perturbation theory. The frequency values are caught by the Random Decrement Technique applied to the time history vibrations for one sensor at the free extremity of the beam. Detection and localization are successful, even for small and transient changes of the structure properties.

11:10-11:30 ThAT2.3

Compressive Sensing for Full Wavefield Image Recovery in Structural Monitoring Applications

Di Ianni, Tommaso	Univ. di Bologna
Perelli, Alessandro	Univ. of Bologna
De Marchi, Luca	Univ. of Bologna
Marzani, Alessandro	Univ. of Bologna

In this paper, a random sampling scheme based on Compressive Sensing (CS) is used in order to reduce the acquisition time of wavefield signals by means of a scanning laser Doppler vibrometer (SLDV) for Structural Health Monitoring (SHM) applications. The sampling process is indeed quite time consuming, because of noise sources and reduced amplitude of acquired signals. By virtue of the sparse characteristic of the wavefield signal representation in terms of sparsity-promoting dictionaries, e.g. Fourier, Curvelet and Wave Atom transforms, the signal can be however recovered through a limited number of measurements. The implemented CS-based procedure has been validated with experimental signals sub-sampled in a pattern of random distributed points, demonstrating the effectiveness of the approach to limit the acquisition time with extremely low information losses.

11:30-11:50 ThAT2.4

Vibration-Based Health Monitoring Approach for Composite Structures Using Multivariate Statistical Analysis

Garcia, David	Univ. of Strathclyde
Trendafilova, Irina	Univ. of Strathclyde
Al-Bugharbee, Hussein	Univ. of Strathclyde

In this paper a novel procedure for damage assessment is suggested which is based on singular spectrum analysis (SSA). The main feature of the method is that it applies Principal Component Analysis (PCA) to the lagged time series, obtained from the measured structural vibration response. In this study the methodology is developed for the case of a free decay response. The measured acceleration vectors are transformed into the frequency domain and then used to define a trajectory matrix. The covariance matrix of the trajectory matrix is decomposed into new variables, the Principal Components (PCs). They define a new space of linearly correlated variables onto which the dynamics/motion of the system can be projected. This decomposition is used to uncover oscillation patterns among other purposes. The method is applied and demonstrated for the case of a simple 2-DoF system. To demonstrate its capabilities for damage diagnosis different levels of stiffness reduction are introduced. The first two PCs are used to visually demonstrate the abilities of the methodology. The Mahalanobis distance is used to develop a classification system to detect and localize delamination in the 2-DoF system. The results clearly demonstrate the capabilities of the system to clearly detect and localize damage.

11:50-12:10 ThAT2.5

Principal Component *Analysis Approach for Sensor Fault Detection and Structure Health Monitoring*

Stoffels, Nicolas	IRSEEM
Sircoulomb, Vincent	IRSEEM
Hermand, Guillaume	ANDRA
Hoblos, Ghaleb	IRSEEM/ESIGELEC

The aim of this paper is to propose an algorithm for detecting faults such as cracks in an underground structure to ensure its health monitoring. The proposed approach is based on the PCA algorithm. Once PCA components are computed, we can see easily the impact of a crack on their norms. The impact represents a good indication to detect abrupt change.

ThAT3 Room I

Achievements and Prospects for Nanotechnologies and Nanosensors in SHM (Invited Session)

Chair: Lebental, Bérengère	IFSTTAR
Co-Chair: Michelis, Fulvio	IFSTTAR

10:30-10:50 ThAT3.1

Self-Powered Conformable Deformation Sensor Exploiting the Collective Piezoelectric *Effect of Self-Organised GaN Nanowires (I)*

Pauliac-Vaujour, Emmanuelle	CEA, LETI, Grenoble, France.
Salomon, Sven	CEA, LETI, Grenoble, France.
Eymery, Joël	CEA, INAC, Grenoble, France.
Leon Perez, Edgar	CEA, LETI, Grenoble, France.

We present a novel integration-driven approach to the design of multi-scale multi-physics sensors and systems. We implement this method to model, design, fabricate and characterize a thin, conformable low-cost impact detection sensor based on assemblies of piezoelectric GaN nanowires. When suitably assembled, the latter demonstrate a macroscale additivity of their nanoscale intrinsic properties, which enables to appeal to classical fabrication techniques and exploitable electronic readouts at the system level. We also exploit multi-level simulations to provide useful insights of adapted application-driven integration solutions for these new forms of sensors. We demonstrate the potential of such application-targeted, fully-integrated and modular systems to accommodate to the stringent requirements of structural health monitoring (SHM).

10:50-11:10 ThAT3.2

Strain Sensors Based *on Carbon Nanotube – Polymer Coatings*

Grabowski, Krzysztof	AGH Univ. of Science and Technology, Kraków
Zbyrad, Paulina	AGH Univ. of Science and Technology, Kraków
Wilmański, Alan	AGH Univ. of Science and Technology, Kraków
Uhl, Tadeusz	AGH Univ. of Science and Technology, Kraków

In this work there have been investigated the potential usage of the CNT's as strain sensors for the structural health monitoring based on the spray coatings. Experimental work was performed on the metal and glass-reinforced composites. Multiwalled Carbon Nanotubes (MWCNTs) were mixed with different matrix materials (acrylic and epoxy) and then applied to the test material with the use of two techniques (screen printing and spray coating).

Futhermore, sensors were investigated using SEM. Response of the sensors was measured due to the tensile test of the specimens..

11:10-11:30 ThAT3.3

*An Innovative Sensor for Weigh-In-*Motion *Applications (I)*

Ghaddab, Boutheina	IFSTTAR
Lebental, Bérengère	IFSTTAR
Gaudefroy, Vincent	IFSTTAR
Michelis, Fulvio	IFSTTAR
Ruiz Hitzky, Eduardo	Materials Science Inst. of Madrid
Aranda, Pilar	Materials Science Inst. of Madrid
Ruiz-Garcıa, Cristina	Materials Science Inst. of Madrid

This study deals with the development of an innovative weigh-in-motion (WIM) sensor. An electrically conductive nanocomposite material based on a mixture of graphene supported on sepiolite and carbon nanotubes was developed. Deposited on bituminous mix with copper electrodes, it is used as a force sensor. We detail the sensor fabrication process and study its sensitivity to a compressive force.

11:30-11:50 ThAT3.4

Wireless Flexible Strain Sensor Based on Carbon Nanotube Piezoresistive Networks for Embedded *Measurement of Strain in Concrete (I)*

Michelis, Fulvio	IFSTTAR
Bodelot, Laurence	Ec. Pol.
Cojocaru, Costel-Sorin	Ec. Pol.
Sorin, Jean-Luc	IFSTTAR
Bonnassieux, Yvan	Ec. Pol.
Lebental, Bérengère	IFSTTAR

This work proposes a new type of low-cost strain sensor, based on piezoresistive carbon nanotube (CNT) network deposited on a flexible substrate. Experimental results show that the strain can be reliably measured thanks to the highly linear piezoresistive behaviour of the CNT network and thanks to temperature compensation capabilities. Moreover, the experimental results show the capability of measuring multiple loading cycles. The performance and the range of sensitivity of the device, suggest possible usage in the domain of embedded monitoring, in particular the detection of micro-strain and micro-cracking in concrete. In order to target this domain, a wireless RFID solution to embed the sensor into concrete is provided.

11:50-12:10 ThAT3.5

Nanocomposite Based Structural Health Monitoring Approaches for Fibre Reinforced Polymers - on Doorstep to Industrial Relevance after Two Decades of Intense Research?

Fritsch, Daniel	Tech. Univ. Hamburg Harburg
Viets, Christian	Tech. Univ. Hamburg Harburg, Inst. of Pol. and
Mannov, Evgenij	Tech. Univ. Hamburg Harburg, Inst. of Pol. and
Schulte, Karl	Tech. Univ. Hamburg Harburg, Inst. of Pol. and
Fiedler, Bodo	Tech. Univ. Hamburg Harburg, Inst. of Pol. and

Reliability and availability thus minimized cost-intensive downtime are mandatory for successful application of fibre reinforced polymer (FRP) structures in many industrial branches. Different approaches for Structural Health Monitoring (SHM) systems compete for industrial application. Nanoscale carbon particle filled polymer matrices exhibit good performance regarding structural integrity as well as spatial resolution to create functionally enhanced composites. Changes of electrical signal due to deformation of the modified polymer are used as corresponding piezoresistive signal ("Sensing"). Major drawbacks like incompatibility of implementation strategies into existing manufacturing processes or insufficient sensor properties anticipate propagating implementation of nanocomposite SHM approaches especially in large structures. Therefore fundamental material's knowledge for property tailoring combined with well-thought-out implementation strategies are necessary. The present contribution offers an overview over today's research on nanocomposite based SHM trials from nanoscale material level to promising implementation strategies.

ThAT4 Room J

Electromagnetic SHM (Regular Session)

Chair: Pierce, Stephen Gareth	Univ. OF STRATHCLYDE
Co-Chair: Ostiguy, Pierre Claude	GAUS – Dept Mechanical Engineering – Univ. of Sherbrooke Sherbrooke (QC), CANADA, J1K2R1

10:30-10:50 ThAT4.1

Hidden Damage Detection for Main Cables of Suspension Bridges Incorporating DC Magnetization *with a Search Coil-Based B-H Loop Measurement*

Kim, Ju-Won	Sungkyunkwan Univ.
Moon, Dae Joong	EJ Tech.
Park, Seunghee	Sungkyunkwan Univ.

Recently, there have been lots of research and development activities to confirm the reliability in terms of the healthy condition for hanger cables of suspension bridges. However, little research about non-destructive evaluation (NDE) techniques for main cable of the suspension bridges has been carried out. In this context, this study proposes a new methodology incorporating DC magnetization with a search coil-based total flux measurement to investigate the loss of the cross section of the main cable due to corrosion damage. From the hysteresis curve of the magnetized main cable measured at the search coil of the proposed NDE equipment, a quantification algorithm to estimate the loss rate of the main cables' cross section is induced. The feasibility of the proposed NDE methodology is verified throughout an experimental study using a real scaled cable specimen with artificially inflicted broken wires.

10:50-11:10 ThAT4.2

Mechanical Vibration Sensing for Structural Health Monitoring Using a Millimeter-Wave Doppler *Radar Sensor*

Moll, Jochen	Goethe Univ. of Frankfurt
Bechtel, Kaspar	Goethe Univ. of Frankfurt
Hils, Bernd	Goethe Univ. of Frankfurt
Krozer, Viktor	Goethe Univ. of Frankfurt

In-service detection of material failures based on an integrated sensor network is the main goal of structural health monitoring (SHM). In this paper, we report on a millimeter-wave Doppler radar sensor operating at 102GHz that enables accurate and non-contact mechanical vibration sensing. Thanks to the unique properties of mm-wave radiation it is possible to penetrate through many non-conducting materials. In contrast to Laser-Doppler-vibrometry this modality enables non-contact vibration measurements behind barriers such as glass-fiber-reinforced plastics, sandwich structures, foams, ceramics etc. Experimental measurements are shown for damage detection and non-linear motion sensing. The unique properties of mm-wave radiation promise great potential for a multitude of future SHM-applications.

11:10-11:30 ThAT4.3

Characterization of an Excavated Pre-Trajan Wall Structure by Integrated Geophysical *Methods*

Orlando, Luciana	Sapienza Univ. of Rome
Cardarelli, Ettore	Univ. of Rome "Sapienza"
Cercato, Michele	Sapienza Univ. of Rome
De Donno, Giorgio	Sapienza Univ. of Rome
Renzi, Beatrice	Sapienza Univ. of Rome

The aim of this study is the characterization of a pre-Trajan wall structure decorated with mosaics and located below the Cryptoporticus of the "Baths of Trajan" complex in Rome. The surveyed wall is 15 m large and 0.9 m wide, with an height ranging from 3 to 5 m. Ground Penetrating Radar (GPR) and Seismic Refraction Tomography profiles were performed on the wall for reconstructing its inner geometry as well as characterizing its building materials, with the additional purposes of mapping fractures and evaluating seismic velocities of materials, to give safety indications before restarting the excavation. The wall was surveyed with horizontal and vertical massive GPR profiles and with two seismic lines. Both the seismic lines and the GPR profiles are able to detect the discontinuity between the two different materials forming the wall, confirming the consistency of the joint interpretation of the two different non-invasive techniques. Moreover they allowed us to locate weak materials and fractures. These results will be used for planning a safety project for the future archaeological excavations. We demonstrate that this approach constitutes an important tool for the characterization of the current status of an archaeological wall.

11:30-11:50 ThAT4.4

Fundamental Study on NDT of Building *Wall Structure by Radar*

Sato, Motoyuki — Tohoku Univ.
Takahashi, Kazunori — Tohoku Univ.
Liu, Hai — Tohoku Univ.
Koyama, Christian N. — Tohoku Univ.

GB-SAR (Ground Based Synthetic Aperture Radar) for inspection of wooden and concrete walls and structures has been developed and the system was evaluated by test measurements. The GB-SAR system uses frequency bandwidth 1-10GHz, and it can acquire full polarimetric radar signal. Then Synthetic Aperture (SAR) Radar signal processing is used to reconstruct 3-dimenatioal images of inner structure of the targets. We found that the radar polarimetry gives us very precise information of the damaged structures, and demonstrated that radar polarimetry is a useful tool for detecting fractures inside a concrete structures, and detection of small deformation of wooden structures.

11:50-12:10 ThAT4.5

Electromechanical Impedance Spectroscopy and Guided Wave Propagation Predictive Modeling on *Composite Materials*

Gresil, Matthieu — Univ. of Manchester
Giurgiutiu, Victor — Univ. of South Carolina

Electromechanical impedance spectroscopy (EMIS) and guided wave (GUW) propagation are a popular structural health monitoring (SHM) technique, which had found applications in many fields of engineering: mechanical, aerospace, civil and others. Piezoelectric wafer active sensors (PWAS) are lightweight and inexpensive transducers that enable a large class of SHM applications such as: (a) embedded GUW ultrasonic, i.e., pitch-catch, pulse-echo, phased arrays; (b) high-frequency modal sensing, i.e., EMIS method; and (c) passive detection, i.e., acoustic emission (AE). The aim of the work presented in this paper is to provide tools to extend modelling capacities and improve quality and reliability of EMIS and 2-D GUW propagation models using commercially available multi-physics finite element method (MP-FEM) packages on fibre reinforces polymers (FRP). The focus of this paper is on the challenges posed by using PWAS transducers in the composite laminate structures as different from the metallic structures on which this methodology was initially developed.

ThAT5 — KL Room
Civil Engineering SHM (Regular Session)

Chair: Faravelli, Lucia — Univ. of Pavia
Co-Chair: Hornych, Pierre — IFSTTAR

10:30-10:50 ThAT5.1

Monitoring of Artificial Defects within a Pavement Structure with a NDT Method Based on a Mechanical *Impact*

Simonin, Jean-Michel — IFSTTAR
Hornych, Pierre — IFSTTAR

This paper presents a non-destructive testing (NDT) method used to monitor a pavement structure which contains artificial defects. A 25 m long pavement section has been built on the full scale accelerated pavement testing facility of IFSTTAR in Nantes. The structure is made of two bituminous layers (8 cm thick base layer, and 6 cm thick wearing course), over a granular subbase. Several types of defects have been included at the interface between the two asphalt layers (figure 1). Rectangular debonded areas of different size (of longitudinal or transversal direction) have been created artificially, using different techniques (sand, textile, absence of tack coat). The construction has been carried out by a road construction company, using standard road works equipment. Then, the pavement fatigue testing facility has been used to apply traffic loading on this pavement, to study the effect of such sliding interfaces on the mechanical behaviour of the pavement, and the evolution of the defects with traffic.

10:50-11:10 ThAT5.2

Autonomous Crack *Measurement for Comparison of Vibratory Compaction Excitation and Climatological Effects*

Dowding, Charles — Northwestern Univ.

Establishment of the fragility of historic structures near rights of way often produces costly delays in construction. Autonomous Crack Measurement (ACM) of micro meter response of cracks can be employed to assess the potential for adjacent construction activities to cause cosmetic cracking in both historic and non-historic structures. ACM measurement of micro meter crack width changes caused by long term climatological and vibratory roller compaction adjacent to an adobe (native mud brick) structure is employed to illustrate the use of micro meter changes in crack width as a tool for structural health monitoring.

It was found that long term or weather-induced crack responses are much larger than those induced by high frequency vibratory compaction, even at US and above European regulatory limits. At more typical levels the potential for formation of cosmetic cracks in newer adobe structures are negligible. Thus for this class of adobe structures, normal regulatory guidance is sufficient for construction or operation of transportation facilities.

11:10-11:30 ThAT5.3

Use of Distributed Fiber *Optic Sensors to Detect Damage in a Pavement*

Blanc, Juliette — IFSTTAR
Chapeleau, Xavier — IFSTTAR
Hornych, Pierre — IFSTTAR

This paper presents the feasibility of damage detection in the asphalt pavements by embedded fiber optics as a new non-destructive inspection technique. The distributed fiber optic sensing technology called "Rayleigh technique" was used in this study. The main advantage of this technique is that it allows to measure strains over long length of fiber optic with a high spatial resolution, less than 1 cm. By comparing strain profiles measured at different time, an attempt was made to link strain changes with the appearance of damage (cracking) in the pavement. This non-destructive method was evaluated on the IFSTTAR accelerated pavement testing facility, in a bituminous pavement. In our experimentation, the optical fibers were placed near the bottom of the asphalt layer. The application of 728 000 heavy vehicle loads (65 KN dual wheel loads) was simulated in the experiment. Optical fiber measurements were made at regular intervals and surface cracking of the pavement was surveyed. After some traffic, a significant increase of strains was detected by the optical fibers at different points in the pavement structure, before any damage was visible. Later, cracking developed in the zones where the strain profiles were modified, thus indicating a clear relationship between the increased strains and crack initiation. These first tests demonstrate that distributed fiber optic sensor based on Rayleigh sensing technique can be used to detect crack initiation and propagation in pavements, by monitoring strain profiles in the bituminous layers.

11:30-11:50 ThAT5.4

Assessing the Risk of Salt Weathering in Sandstone by Instrumented Monitoring *of the Electrical Impedance*

Lehmann, Frank — Materialprüfungsanstalt Univ. Stuttgart

Krüger, Markus	Materialprüfungsanstalt Univ. Stuttgart

Salt weathering of natural stone is one of the most severe and most costly deterioration at historic buildings. The existence of salts and salt mixtures within the material requires adequate measures to avoid further deterioration. An inevitable prerequisite before conservation activities can be conducted is the preliminary assessment by laboratory investigations, but also by increasing the knowledge of the environmental influences that affect the salt weathering rate and the actual processes in the stone. The instrumented monitoring of a material's electrical impedance in combination with the ambient conditions allows the acquisition of information on its moisture content, both with and without salts. A joint analysis including the supplemental measured parameters may be used to draw conclusions on the dynamic interaction between moisture and salt transport, as well as salt phase changes, and hence provides the possibility to evaluate the risk of salt weathering to the structure.

11:50-12:10 ThAT5.5

Damage Detection Methods on Wind Turbine Blade Testing with Wireless Accelerometer Sensors

Mollineaux, Mark	Stanford Univ.
Balafas, Konstantinos	Stanford Univ.
Branner, Kim	Department of Wind Energy, Tech. Univ. of Denmark
Kiremidjian, Anne	Stanford
Rajagopal, Ram	Stanford Univ.
Tesauro, Angelo	Tech. Univ. of Denmark
Nielsen, Per Hørlyk	Tech. Univ. of Denmark

Testing was performed on a 34 meter blade at a facility in DTU Risø Campus, featuring both wired accelerometers and low-power MEMs-based wireless accelerometers. Testing was focused on an induced delamination area on the trailing edge of the blade, which was subject to various configurations in order to simulate different degrees of damage. Excitation was performed in two ways: near the delamination zone in a simulation of operational wind excitations, and with a bar designed to excite torsional modes of the wind turbine blade. We compare the data collected from the wireless sensors against wired sensors to demonstrate their performance. We explore methods for determining damage. We first explore results of autoregressive coefficients for indicating damage levels. Finally, we demonstrate the use of damage sensitive features from the wavelet transforms of input and output signals to provide a method suitable for non-stationary blade excitations.

ThAT6 Room 200

Damage I (Regular Session)

Chair: Ni, Yi-Qing	The Hong Kong Pol. Univ. Hong Kong
Co-Chair: Wu, Wen-Hwa	National Yunlin Univ. of Science and Tech.

10:30-10:50 ThAT6.1

Guided Wave Propagation through Composite Bonded Joints

Sherafat, Mohammad Hossein	McGill Univ.
Quaegebeur, Nicolas	GAUS - Univ. of Sherbrooke
Lessard, Larry	McGill Univ.
Hubert, Pascal	McGill Univ.
Masson, Patrice	GAUS – Dept Mechanical Engineering – Univ. of Sherbrooke Sh

Bonding of composite structures is regularly used in aerospace industry to restore structural integrity in repair or assemble sub-structures. This structural feature is prone to degradation with improper installation or disbond when submitted to fatigue or extreme strains. Therefore, non-destructive evaluation (NDE) should be considered to inspect adhesive bonded joints. In comparision with traditional NDE, a built-in structural health monitoring (SHM) would be more desirable to save the inspection cost as well as improve safety. The objective of this research is to evaluate integrity of composite bonded joints using guided wave propagation technique. Two carbon fiber reinforced polymer (CFRP) are bonded together using adhesive film and artificial disbond is introduced by a circular Teflon tape. A rectangular piezoceramic is used to generate plane guided waves and non-contact measurement is performed using 3-D Laser Doppler Vibrometer to extract the required information for evaluation of quality of bond line. Results have shown that the joint can be charactized using guided wave behavior; and reflection and transmision coefficients for incident A0 and S0 modes are extracted in order to design the final SHM system.

10:50-11:10 ThAT6.2

Nonlinear Elastic Tomography *Using Sparse Array Measurements*

Ciampa, Francesco	Univ. of Bath
Meo, Michele	Univ. of Bath
Scarselli, Gennaro	Univ. del Salento, Lecce, Italy
Pickering, Simon	Univ. of Bath
Giampiccolo, Andrea	Univ. of Bath

Literature offers a quantitative number of diagnostic imaging methods that can continuously provide a detailed image of the material defects in aerospace and civil application. This paper presents a nonlinear Structural Health Monitoring (SHM) imaging method, based on nonlinear elastic wave tomography (NEWT), for the detection of the nonlinear signature in damaged isotropic structures. The proposed technique, based on a combination of higher order statistics (HOS) and radial basis function (RBF) interpolation, is applied to a number of waveforms containing the nonlinear responses of the medium. HOS such as bispectral analysis and bicoherence was used to characterize the second order nonlinearity of the structure due to corrosion, whilst RBF interpolation was applied to a number of signals acquired from a sparse array of sensors, in order to obtain an image of the defect. Compared to standard linear ultrasonic imaging techniques, the robustness of this nonlinear tomography sensing system was experimentally demonstrated. Moreover, this methodology does not require any baseline with the undamaged structure for the detection of the nonlinear source as well as a priori knowledge of the mechanical properties of the medium. Finally, the use of HOS makes NEWT a valid alternative to traditional nonlinear elastic wave spectroscopy (NEWS) methods for materials showing either classical or non-classical nonlinear behaviour.

11:10-11:30 ThAT6.3

Onboard – SHM System Using Fibre Optical Sensor and LAMB Wave Technology for Life Time Prediction and Damage Detection on Aircraft Structure

Berger, Ulrich	IABG mbH
Hayo, Tim	IABG mbH

Structural Health Monitoring (SHM) is often understood as the monitoring of the structure's fatigue. The fatigue consumption of an aeronautic structure is performed either by load or respectively strain monitoring and the analysis of the strain spectra or by identifying the fatigue crack initiation in-service. The fatigue life estimation in such cases is only possible with respect to the macroscopic crack growth. This paper is split in two parts. On one hand, it will give an overview about the IABG activities regarding the development and aeronautic application of SHM systems. Therein, it is demonstrated that a combined usage of fibre optical sensors as well as Lamb wave analysis on the fatigue test of an aircraft structural component allows precise conclusion regarding fatigue life consumption. It further allows to benchmark both technologies. Therefore, a specific concept and development of an on-board SHM System for the in-service use on an aircraft is presented. On the other hand, the requirements in order to technically qualify a generic SHM-system are illustrated and how IABG can support the SHM developer in this process, especially when higher requirements of Technology Readiness Levels are to be validated by the developer.

11:30-11:50 ThAT6.4

Voting Neural Network Classifier *for Detection of Fatigue Damage*

in Aircrafts

Dworakowski, Ziemowit	AGH Univ. of Science and Technology, Kraków
Ambrozinski, Lukasz	AGH Univ. of Science and Technology, Kraków
Dragan, Krzysztof	Air Force Inst. of Tech.
Stepinski, Tadeusz	AGH Univ. of Science and Technology, Kraków
Uhl, Tadeusz	AGH Univ. of Science and Technology, Kraków

An ANN based method for detection and localization of fatigue damage in aircraft structures is presented in the paper. Damage indices are calculated from Lamb-wave measurements conducted by the network of piezoelectric transducers. Data gathered by the sensors is used as an input to the proposed voting neural network classifier. A set of neural network electors of different architecture cooperates to achieve consensus concerning the state of each monitored path. Sensed signal variations in the ROI, detected by the networks at each path, are used to assess the state of the structure as well as to localize detected damage and to filter out ambient changes. The classifier has been extensively tested on large data sets acquired in the tests of specimens with artificially introduced notches as well as the results of numerous fatigue experiments. Effect of the classifier structure and test data used for training on the results is evaluated. It is shown that the developed classifier performs better than individual ANNs in terms of damage detection. The classifier structure, composed of different networks working together, yields an increased reliability, mainly due to the lower impact of the initial weights distribution on the final result.

11:50-12:10 ThAT6.5

Pitch-Catch Air-Coupled Ultrasonic Technique for Detection of Barely Visible Impact Damages *in Composite Laminates*

Ambrozinski, Lukasz	AGH Univ. of Science and Technology, Kraków
Piwakowski, Bogdan	Ec. Centrale de Lille
Stepinski, Tadeusz	AGH Univ. of Science and Technology, Kraków
Pieczonka, Lukasz	AGH Univ. of Science and Technology, Kraków
Uhl, Tadeusz	AGH Univ. of Science and Technology, Kraków

In this paper we demonstrate an application of Lamb waves, evoked and sensed using air-coupled transducers, with the aim to detect barely visible impact damages (BVID). Experimental results, performed using a pair of air-coupled transducers operating in pitch-catch setup, are presented. In this setup, selective excitation and reception of selected Lamb wave modes is possible by changing transducers' angle of incidence. Results of scanning of a carbon-fiber reinforced polymer (CFRP) panel using S0 Lamb wave mode are presented in the paper, and different damage detection strategies are analyzed to detect impact damages in the panel.

ThBT1 Auditorium

Strain Monitoring II (Regular Session)

Chair: Lecieux, Yann	GeM, Univ. of Nantes
Co-Chair: Casari, Pascal	Univ. of Nantes

13:50-14:10 ThBT1.1

Robust Virtual Dynamic Strain *Sensors from Acceleration Measurements*

Tondreau, Gilles	Univ. libre de Bruxelles (ULB)
Deraemaeker, Arnaud	Univ. Libre de Bruxelles

The possibility to deduce dynamic strains from acceleration measurements is investigated in this paper. The classic technique applied on beam-like structures which consists in using vertical accelerations to estimate longitudinal strains with a second order central finite difference is compared with two methods based on horizontal accelerations. Time domain responses of accelerometers and strain sensors installed on a simply supported beam are simulated, and the trains deduced from accelerations are compared with the real strains. The possibility of locating damage with the mode shape obtained with estimated strains is also addressed.

14:10-14:30 ThBT1.2

Long-Term Monitoring *of High-Rise Buildings in Moscow*

Kapustian, Natalia	Schmidt Inst. of Physics of the Earth of the Russian Acad.
Voznyuk, Alexander	Central Res. and Design Inst. for Res. and Public
Klimov, Alexey	Moscow State Univ. of Civil Engineering, Central Res. a

A long-term structural health monitoring program started in 2000 with a seismometrical monitoring station installed in 44-storey "Edelweiss" tower. Since then several high-rise residential buildings in Moscow were equipped with monitoring systems. The research includes both practical and scientific purposes. The practical one is to increase the structural safety by detecting anomalies in structural behaviour. The scientific one is to gain knowledge about the evolution of reinforced concrete structures stress-strain state during the construction and operation of buildings. The large amounts of data have been obtained using various types of equipment (seismometers, pressure cells, extensometers and vibrating wire strain gauges). Based on results it was possible to establish main factors that influence the stress-strain state of structures. The actual data was compared to simulation results and the most appropriate computer models were determined. The redistribution of stresses between parts of a construction turned out to be impossible to reproduce by means of finite element analysis in current software suites was found. Finally, a technique of long-term forecast of the stress-strain state of reinforced concrete structures was developed.

14:30-14:50 ThBT1.3

An Ultrasonic *Strain Gauge*

Kersemans, Mathias	Ghent Univ.

A method is introduced for the measurements of strain exploiting the interaction between ultrasound waves and characteristics of the insonified specimen. First, the response of obliquely incident harmonic waves to a deterministic surface roughness is utilized. Analysis of backscattered amplitudes in Bragg diffraction geometry then yields a measure for the in-plane strain field by mapping any shift in angular dependency. Secondly, the analysis of the reflection characteristics of normal incidence pulsed waves in frequency domain provides the out-of-plane normal strain field component, simply by tracking any change in the stimulation condition for a thickness resonance. As such, the developed method yields an absolute, contactless and single-sided mapping of a local 3D strain field, in which both sample preparation and alignment procedure are ubiquitous. The ultrasonic strain gauge is applicable to any material, though under the restriction that a deterministic surface roughness is present. Results are presented for cold-rolled DC06 steel samples onto which skin passing of the work rolls is applied. The samples have been mechanically loaded, introducing plastic strain levels ranging from 2% to 35%. The ultrasonically measured strains have been validated with other strain measurement techniques, yielding good agreement. As the ultrasonic strain gauge provides all three strain field components, we extracted Lankford ratios for the DC06 steel sheet at different applied plastic strain levels revealing a strain dependent plastic anisotropy of the DC06 steel.

14:50-15:10 ThBT1.4

Aircraft Parametric Structural Load Monitoring Using Gaussian Process Regression

Fuentes, Ramon	Univ. of Sheffield
Cross, Elizabeth	Univ. of Sheffield
Worden, Keith	Univ. of Sheffield
Halfpenny, Andrew	HBM-nCode
Barthorpe, Robert J.	Univ. of Sheffield

The work presented here demonstrates the capability of Gaussian Process (GP) regression for the prediction of aircraft structural loads based on recorded flight parameters. The objective of monitoring aircraft loads during operation is to develop a better understanding of aircraft usage and thus to provide the operator with an accurate estimate of the remaining useful life of components against their prescribed fatigue life. These loads are often difficult and expensive to measure and this motivates the use of advanced mathematical techniques to estimate them accurately based on other parameters that are typically measured during flight. Gaussian Process regression is a powerful Bayesian machine learning tool whereby predictions and their distributions can be obtained without having to specify a particular model/functional form. Data collected from a military trainer aircraft is used to demonstrate how the mapping of measured strains to basic flight parameters such as airspeed, accelerations, and control surface deflections can be performed using GP regression. It is also shown how these results can be applied in an aircraft usage monitoring context. The GP predictions for strain are compared to the actual measurements for 101 flights, and the results are presented in terms of fatigue life, correlation, and mean-squared error. The results are encouraging, with errors in fatigue life in the range of 5% to 30% in the worst cases.

15:10-15:30 ThBT1.5

Quantitative Strain Measurement with Distributed Fiber Optic Systems: Qualification of a Sensing *Cable Bonded to the Surface of a Concrete Structure*

Billon, Astrid	EDF R&D
Henault, Jean-Marie	EDF R&D
Quiertant, Marc	Univ. Paris-Est - IFSTTAR
Taillade, Frédéric	EDF R&D
Khadour, Aghiad	Univ. Paris-Est - IFSTTAR
Martin, Renaud-Pierre	Univ. Paris-Est - IFSTTAR
Benzarti, Karim	Univ. Paris-Est - IFSTTAR

Distributed Fiber Optic Systems (DOFS) are an emerging and innovative technology that allows long-range and continuous strain/temperature monitoring with a high resolution. However, strain profiles measured in the optical fiber may differ from actual strain in the structure, due to the shear transfer through the intermediate material layers between the optical fiber and the host material (i.e., in the protective coating of the sensing cable and in the adhesive). Therefore, fiber optic sensors need to be qualified in order to provide accurate quantitative strain measurements. This study presents a methodology for the qualification of a DOFS. It is proposed to establish a numerical modeling of the system, in which the mechanical parameters are calibrated from experiments. A specific surface-mounted sensing cable connected to an Optical Frequency Reflectometry Domain (OFDR) interrogator is considered as case study. We found that (i) tensile and pull-out tests are able to provide full information about materials and interfaces of our modeling; (ii) the calibrated model allows us to compute strain profiles along the optical fiber, (iii) which proved to be consistent with strain profiles measured on a cracked concrete beam during a 4-points bending test.

ThBT2 GH Room

Statistical Approaches II (Regular Session)

Chair: Chatzi, Eleni	Inst. of Structural Engineering, ETH Zurich
Co-Chair: Lanata, Francesca	Ec. Supérieure du Bois

13:50-14:10 ThBT2.1

Gearbox Condition Monitoring *under Variable Operational Conditions*

Toshkova, Daniela	Univ. of Bristol, Beran Inst. Ltd.
Lieven, Nicholas	Univ. of Bristol
Morrish, Peter	Beran Inst. Ltd.
Titurus, Branislav	Univ. of Bristol
Moore, Will	Beran Inst. Ltd.
Brinkworth, Niel	E.ON

The gearbox vibration spectrum is complicated and overloaded with artefacts, making it difficult for the eye to register a change due to a fault. In order to overcome this difficulty this paper presents a gearbox condition indicator, which can be trended. The gearbox condition indicator is based on the summation of the sideband levels around the first three harmonics of the gear mesh frequency. The sensitivity of the proposed gearbox condition indicator to damage is investigated using seeded improper backlash under a range of operational conditions.

14:10-14:30 ThBT2.2

Using Non Continuous Records *from Full Scale Monitoring System for Fatigue Assessment*

Flamand, Olivier	CSTB
De Oliveira, Fabrice	CSTB
Stathopoulos-Vlamis, Aris	GEFYRA
Papanikolas, Panayotis	GEFYRA
Panagis, Akis	GEFYRA

Monitoring systems are prone to record huge amount of data, only a minor part of which could be of interest. Maintaining a giant data base from the monitoring system of a large structure equipped with dozens of sensors is a costly challenge. One way for solving this question lies in discarding everything that does not enter in the scope of what was considered as "interesting event" at the design stage, with a big risk of missing an unattended but essential event. Another way could be to store less data, making a kind of "survey" instead of recording continuously, with a possibility to rebuild events from the discontinuous records. This way of doing is suitable in the case of minutes lasting phenomena only, it is clear that a sudden and brief phenomenon could be missed in such a configuration. This method was used in the case of vortex shedding excitation of the deck of a large cable stayed bridge, in order to rebuild the history

14:30-14:50 ThBT2.3

Examination of Two Sensor Placements Schemes in Damage Detection

Bernal, Dionisio	Northeastern Univ.
Döhler, Michael	Inria
Parker, Danny	AVNIK

Deciding on the position of sensors by optimizing the utility of the monitoring system over a structure's lifetime is typically forbidden by computational cost. Sensor placement strategies are, instead, usually formulated for a pre-selected number of sensors and are based on cost functions that can be evaluated for any arrangement without the need for simulations. This paper examines the performance of two such schemes, the first one is derived directly from a technique that detects damage from the shift of a chi-square distribution from central to non-central and takes the optimal arrangement as the one that maximizes the sensitivity of the non-centrality to all parameter changes of equal norm. The second scheme selects the sensor arrangement as that which maximizes a weighted version of the norm of the sensitivity of the covariance of the output to all feasible changes in system parameters. The performance of the two schemes is tested in simulations.

14:50-15:10 ThBT2.4

Structural Damage Identification *and Prognosis Via a Statistical Learning Approach*

Mao, Zhu	Univ. of California, San Diego
Todd, Michael	Univ. of California, San Diego

Since all damage identification strategies inevitably involve uncertainties from various sources, a higher level of characterization is necessary to facilitate decision-making in a statistically confident sense. Machine learning plays an important role in the decision-making process of damage detection, classification, and prognosis, which employs training data (or a validated model) and extracts useful information from the high-dimensional observations. This paper classifies the type of damage via support vector machine (SVM) in a supervised learning fashion, and selects the most plausible model for data interpretation. Therefore the separation of damage type and failure trajectory is transformed into a group classification process, under the influence of uncertainty. Given data observation, SVM is obtained under a training process, which characterizes the best classification boundaries for any future feature set. A rotary machine test-bed is employed, and vibration-based damage features are evaluated to demonstrate the proposed classification process.

15:10-15:30 ThBT2.5

Optimal Sensors Placement *to Enhance Damage Detection in Composite Plates*

Fendzi, Claude	Arts et Métiers Paristech, PIMM Lab.
Morel, Julien	Arts et Métiers Paristech, PIMM Lab.
Rébillat, Marc	PIMM, Arts et Métiers ParisTech
Guskov, Mikhail	Arts et Métiers Paristech, PIMM Lab.
Mechbal, Nazih	Arts et métiers ParisTech

This paper examines an important challenge in ultrasonic structural health monitoring (SHM), which is the problem of the optimal placement of sensors in order to accurately detect and localize damages. The goal of this study is to enhance damage detection through an optimal sensor placement (OSP) algorithm. The problem is formulated as a global optimization problem, where the objective function to be maximized is evaluated by a ray tracing approach, which approximately models Lamb waves propagation. A genetic algorithm (GA) is then used to solve this optimization problem. Simulations and experiments were conducted to validate the proposed method on a carbon epoxy composite plate. Results show the effectiveness and the advantages of the proposed method as a tool for OSP with reasonable computation time.

ThBT3 Room I

SHM Applications II (Regular Session)

Chair: Ostachowicz, Wiesław	Pol. Acad. of Sciences, Inst. of Fluid Flow Machinery
Co-Chair: Simonin, Jean-Michel	IFSTTAR

13:50-14:10 ThBT3.1

An Experimental Study of the Pseudo-Local Flexibility Method for Damage Detection of Hyper-*Static Beams*

Hsu, Ting Yu	National Center for Res. on Earthquake Engineering
Shiao, Shen-Yuan	National Taipei Univ. of Tech.
Liao, Wen-I	National Taipei Univ. of Tech.
Loh, Chin-Hsiung	National Taiwan Univ.

Many vibration-based structural damage detection techniques which perform damage diagnosis of a structure based on structural dynamic characteristic parameters have been proposed in the last two decades. One of the promising approaches proposed recently is the local flexibility method. The local flexibility method, which is founded on virtual forces that cause nonzero stresses in a local part of the structure, can estimate damage locations and local stiffness variations of beam structures. It does not require a finite element model of the beam structure. The structural modal parameters identified from the ambient vibration signals both before and after damage are the key information for the local flexibility method. The number of modes necessary for the local flexibility method is usually quite small, especially for a simply supported beam where only the first mode could be sufficient. However, for a hyperstatic beam, the number of modes required for estimation of the damage could be much higher. This makes the feasibility of the local flexibility method much lower because in practice only the first few modes could be identified with high quality using ambient vibration signals. Therefore, in this study, non-local virtual forces which cause concentrated stresses in a local part and nonzero stresses in the other parts of a structure are employed. The theoretical basis of the proposed method which uses non-local virtual forces is derived. The proposed method is validated with a continuous steel beam experiment. The results illustrate that the non-local virtual forces can determine the local variations of stiffness more accurately with less identified modes. Therefore, the feasibility of the proposed method is higher because limited number of high quality modes can be iden

14:10-14:30 ThBT3.2

Temperature and Load Effects in Modeling and Experimental Verification of Acoustic Emission Signals for Structural Health Monitoring Applications

Maurice, Boon	Delft Univ. of Tech.
Zarouchas, Dimitrios	Delft Univ. of Tech.
Martinez, Marcias	Delft Univ. of Tech.
Gagar, Daniel	Cranfield Univ.
Rinze, Benediicus	Delft Univ. of Tech.
Foote, Peter	Cranfield Univ.

The present study focuses on understanding the effect of load and temperature on Acoustic Emission (AE) signal propagation in an Aluminium 2024-T3 panel. In addition, the ability of an AE system to locate damage under these operational and environmental conditions was evaluated. The work was performed in two stages. In stage one, the wave group velocities of guided Lamb waves were measured for a range of temperatures from -40℃ to +70℃. At each temperature level, six different static loads were applied that ranged from 0 MPa to 250 MPa in increments of 50 MPa. It was observed that the variation of temperature and load altered the wave group velocities, which were verified with analytical solutions found in the literature. In stage two, a representative AE signal, simulating a fracture phenomenon was emitted from a randomly selected point. Using values of wave velocity measured in stage one, the location of the representative AE signal under these conditions was calculated and errors were determined. It was found that the location algorithm was not sensitive to wave group velocities changes due to temperature and loads, thus providing an accurate location of the source within 1cm for 93% of the cases studied.

14:30-14:50 ThBT3.3

Approach to Health Monitoring of an Aircraft Structure with Resistive Ladder Sensors During Full Scale Fatigue Test

Artur, Kurnyta	Air Force Inst. of Tech.
Dziendzikowski, Michal	Air Force Inst. of Tech.
Dragan, Krzysztof	Air Force Inst. of Tech.
Leski, Andrzej	Air Force Inst. of Tech.

This paper presents an application of resistive ladder sensors for health monitoring of an aircraft structure. Cracks forming can affect significantly the integrity thus the use of methods which enable damage detection as early as possible is an important issue. The article presents novelty in approach to detect fatigue cracks in the early stage with use of Resistive Ladder Sensors. Additionally, measurement results for customized sensor are delivered, which is design and manufactured on site with conductive paint in DW (direct-write) technique. Moreover the article will present on-line diagnostics capability for the real aircraft elements (including shape complicated) as well as results of the damage detection from full scale fatigue test of turbo-prop military trainer.

14:50-15:10 ThBT3.4

Smart Textile Reinforced *Concrete Sensory Structures*

Goldfeld, Yiska	Tech. - Israel Inst. of Tech.
Rabinovitch, Oded	Tech. - Israel Inst. of Tech.
Quadflieg, Till	Inst. fuer Textiltechnik, RWTH Aachen Univ.
Fishbain, Barak	Tech. –Israel Inst. of Tech.
Gries, Thomas	Inst. fuer Textiltechnik, RWTH Aachen Univ.

This study examines and demonstrates the feasibility of a new class of smart textile reinforced concrete (TRC) structural elements with inherent sensing capabilities that are based on embedding metallic yarns in the textile mesh. The new approach combines the advantages of thin walled glass fiber based TRC with the electro-mechanical properties of the stainless steel fibers embedded in the textile matrix. To examine this concept and to demonstrate its potential feasibility, TRC beam specimens are tested and monitored under mechanical and environmental loading condition. The results of the tests demonstrate the features of the sensory/structural system, reveals its potential use as a basis for a combined structural and functional monitoring system, and highlights its spectrum of potential applications.

15:10-15:30 ThBT3.5

Monitoring of the Cracks *on Bell Tower of S. Anastasia Cathedral in Zadar Croatia*

Uglešić, Davor — D&Z doo
Bohinc, Uros — zag (www.zag.si)

The bell tower of the cathedral of St. Anastasia in Zadar is a major landmark and an important cultural object which has been a subject of strong restoration and preservation efforts for many years. In order to determine the long term stability of the existing cracks, a monitoring system consisting of 8 fiber optic displacement sensors was installed in October 2012. This paper presents the previous structural assessment work as well as the results of the crack displacement measurements. Special attention is given to the analysis of correlation of crack displacements to the temperature and possible other outer disturbances.

ThBT4 Room J
Guided Waves IV (Regular Session)

Chair: Giurgiutiu, Victor — Univ. of South Carolina
Co-Chair: Alleyne, David — Guided Ultrasonics Ltd

13:50-14:10 ThBT4.1

*Crack Detection in Multi-*Wire *Cables Using Guided Ultrasonic Waves*

Schaal, Christoph — Univ. of Stuttgart
Gaul, Lothar — Univ. of Stuttgart, Inst. of Applied and Experimental M

Automated Structural Health Monitoring schemes are developed to cost-efficiently prevent failure of mechanical and civil structures, and to predict the structure's residual life.

In this work, an efficient crack detection algorithm based on the Hilbert transform is presented. By means of this algorithm, crack localization in multi-wire cables is performed through a time-of-flight analysis of the wave packets. Crack identification can be performed by evaluating the waves' amplitudes. The algorithm is fully automatized and distinguishes between wave packets from different waves independently. Its applicability is analyzed for a single cylindrical wire and for multi-wire cables.

14:10-14:30 ThBT4.2

Modal Parameter Determination *of Stay Cable with an Improved Algorithm Based on Stochastic Subspace Identification*

Wu, Wen-Hwa — Nation Univ. of Science and Tech.
Chen, Chien-Chou — Department of Construction Engineering, National Yunlin Univ.
Wang, Shen-Wei — National Yunlin Unisersity of Science and Tech.
Lai, Gwolong — National Yunlin Unisersity of Science and Tech.

The modal parameter determination of stay cable is generally a critical issue for either designing or monitoring cable-stayed bridges. Stochastic subspace identification (SSI) has been proved in recent years to be an excellent tool for obtaining all the modal parameters together in the cases where only the output signals are available. Nevertheless, its feasibility and accuracy in practical applications remain to be extensively verified. Especially for the situations where the frequency content of excitation is narrowly banded and thus far away from the white-noise assumption underneath the theoretical derivation of SSI, the problem of superfluously indentifying numerous deceptive modes usually occurs. Attempting to crack this difficulty, an improved algorithm based on SSI is developed in the current work for performing the modal parameter determination of stay cables. This study adopts the covariance type of SSI, usually with better stability and computation efficiency, to establish an effective methodology for extensively identifying the modal parameters of stay cables. Several details of choosing the computational parameters in performing SSI are first discussed, followed by proposing an alternative stabilization diagram such that most modal parameters of cable can stand out and then imposing appropriate sifting criteria to extract the most reliable modal parameters.

14:30-14:50 ThBT4.3

Wave Propagation in *Helical Multi-Wire Cables*

Treyssède, Fabien — IFSTTAR
Laguerre, Laurent — IFSTTAR

Elastic guided waves are of great interest for the inspection of elongated structures. In practice, analytical or numerical modeling tools are required for a better understanding of the propagation of guided waves, which are multimodal and dispersive, and thereby for the optimization of inspection systems. However, cables are complex structures, helical, multi-wired and highly prestressed. This further complicates the interpretation of measurement. This paper is devoted to the modeling of wave propagation inside seven-wire strands, typically encountered in civil-engineering cables. It gives an overview of recent works, mainly conducted at Ifsttar, in order to account for the helical geometry, interwire coupling and prestress in numerical models. Then, the energy transfer from the central wire to the peripheral ones is investigated by considering an excitation localized into the central wire. The numerical results allow to understand how the energy transfer can decrease with frequency as well as to discover a new compressional mode, of local type, which could be of interest for the non-destructive evaluation or the structural health monitoring of cables.

14:50-15:10 ThBT4.4

Damage Detection on a Cable Stayed Bridge Using Wave Propagation Analysis

Quintana Rodríguez, Juan Antonio — Inst. Mexicano del Transporte
Carrión Viramontes, Francisco Javier — Inst. Mexicano del Transporte
Crespo Sánchez, Saúl Enrique — Inst. Mexicano del Transporte

The Global Search Method (GSM) is being proposed for global evaluation of complex structures. This method is based on the analysis of the flexural wave propagation and the comparison of the real structure responses to those from a calibrated simulation model representing the non-damaged condition. Then, the differences are associated to changes in the structural parameters (stiffness, damping, mass, and geometry) or the boundary conditions, and related to a specific region or location within the structure to locate damage and quantify its overall structural effect. The effectiveness of the GSM was evaluated on a cable stayed bridge where damage was simulated when a cable was removed during rehabilitation; dynamic tests were done using a Falling Weight Deflectometer to control the dynamic impact load and previously analogous tests were done to calibrate the finite element model. Results show that the GSM has the ability to identify and quantify damage, where sensitivity increases when the sensors are closer to damage. In the bridge case, damage could not be detected when the distance of the sensors was greater than 65 meters from damage because of signal attenuation.

15:10-15:30 ThBT4.5

Ultrasonic Verification *of Composite Structures*

Pelt, Maurice Mathieu Joseph Marie — Amsterdam Univ. of Applied Sciences
de Boer, Robert Jan — Amsterdam Univ. of Applied Sciences

Sprik, Rudolf — Univ. of Amsterdam
Schoemaker, Christiaan — Univ. of Amsterdam

Ultrasonic Verification is a new method for the monitoring large surface areas of CFRP by ultrasound with few sensors. The echo response of a transmitted pulse through the structure is compared with the response of an earlier obtained reference signal to calculate a fidelity parameter. A change in fidelity over time is indicative for a new defect in the structure. This paper presents an experimental assessment of the effectiveness and reproducibility of the method.

ThBT5 — KL Room
Structural Health Monitoring of Civil Engineering Structures (Invited Session)

Chair: Bourquin, Frédéric — IFSTTAR
Co-Chair: Gueguen, Philippe — ISTerre/UGA/CNRS/IFSTTAR

13:50-14:10 — ThBT5.1

Structural Health Monitoring *of Unique Structures in Abu Dhabi Emirate (I)*

Skolnik, Derek — Kinemetrics
Ciudad-Real, Mauricio — Kinemetrics Open Systems & Services
Kaya, Yavuz — The Univ. of British Columbia
Safak, Erdal — Bogazici Univ. Kandilli Observatory and Earthquake Res.

The term "Structural Health Monitoring (SHM)" refers to the real-time monitoring and alerting of structures state-of-health. The primary objective of SHM is to improve safety and reliability of civil infrastructure by detecting damage before it reaches a critical state and to enable rapid post-event (e.g., earthquake) assessment. In addition, the data from afforded SHM systems are used to validate design assumptions and codes, to predict response for large excitations, to calibrate analytical models, and to develop instantaneous damage and loss maps. As part of the Emirate of Abu Dhabi Seismic Hazard and Risk Assessment (ADSHRA) project, several unique and prestigious structures in Abu Dhabi were installed with state-of-the-art SHM systems. Typical systems are composed of up to 30 accelerometers within the building, a wind velocity/direction sensor at the roof, and a three-component downhole accelerometer near the building footprint. Data from these systems are all time-synchronized and recorded continuously at 200sps in real-time. A real-time data processing and analysis software package is developed to observe and display the dynamic characteristics (e.g., modal frequencies, damping ratios, and mode shapes) and responses (e.g., accelerations, velocities, displacements, and inter-story drifts) of the structures and their time variations. Since most of the SHM data are due to ambient forces (i.e., low amplitude vibrations with very low signal-to-noise ratios), advanced signal processing and system identification techniques, based on statistical signal processing and stochastic filtering theories, are used for data processing and analysis. In April 2013, two large earthquakes struck the region of southern Iran. Although very far away and producing seemingly very low levels of shaking, both events resulted in mass evacuations across many cities throughout the Gulf countries including Abu Dhabi. One obvious explanation for the understandable widespread reaction is that the region is simply not use to seismic activity. However, analyses of data recorded in several tall Abu Dhabi buildings during these events provides additional unique insight into human perception and structural response to prolonged vibrations of

14:10-14:30 — ThBT5.2

Frequency and Damping Wandering *in Existing Buildings Using the Random Decrement Technique (I)*

Gueguen, Philippe — ISTerre/UGA/CNRS/IFSTTAR
Langlais, Mickael — ISTerre/UGA/CNRS
Roux, Philippe — ISTerre/UGA/CNRS
Schinkmann, Jakub — ISTerre/UGA/CNRS
Douste-Bacqué, Isabelle — ISTerre/UGA/CNRS/IFSTTAR

With the increasing use of permanent, continuous and real-time networks, ambient vibrations can provide a simple tool for the identification of dynamic building parameters. This study is focused on the long-term variation of frequency and damping in several buildings, using the Random Decrement Technique (RDT). RDT provides a fast, robust and accurate long-term analysis and improves the reliability of frequency and damping measurements for structural health monitoring. This reveals particularly useful information in finding out precisely how far changes in modal parameters can be related to changes in physical properties. This paper highlights the reversible changes of the structure's dynamic parameters, correlated with external forces, such as temperature and exposure to the sun. Contrasting behaviours are observed, including correlation and anti-correlation with temperature variations.

14:30-14:50 — ThBT5.3

Application of Subspace Identification to the Assessment of Long-Term Monitoring of Structural *Seismic Responses (I)*

Loh, Chin-Hsiung — National Taiwan Univ.
Chao, Shu-Hsien — National Center for Res. on Earthquake Engineering

This paper presents the identification and assessment of the dynamic characteristics of building and bridge structures from the long-term monitoring of seismic response data. Over 40 events of the seismic response data of these structures during the past 10 years were used. To extract the dynamic characteristics of the structures from its seismic response data, the time-domain data-driven subspace identification (SI-DATA) technique is used to extract the system natural frequencies, damping ratios and mode shapes. To enhance the stability of the identified real system poles in using SI-DATA, an algorithm of removing spurious modes is used. Discussion on the identified system dynamic characteristics of the structures under different intensity level of ground excitation is presented. Effect of soil-structural interaction on the identified system natural frequencies and damping ratios by considering either free-field data or basement data as input in subspace identification is also discussed.

14:50-15:10 — ThBT5.4

Evaluation of the Environmental *Effects on a Medium Rise Building (I)*

Boroschek, Ruben — Univ. of Chile
Tamayo, Felipe — Univ. of Chile
Aguilar, Rafael — PONTIFICIA Univ. CATOLICA DEL PERU - PUCP

This article presents the variation of the dynamics properties due to environmental effects of the Central Tower at the Faculty of Physical and Mathematical Science of the University of Chile. This structure is a nine story, shear wall reinforced concrete building, 30 meters high, which has been monitored automatically and remotely since 2009 with a network of 8 uniaxial accelerometers and 9 environmental sensors. The network registers the ambient conditions, such as wind speed and direction, temperature, radiation, rainfall, ambient and soil humidity. The range of variation on environmental effect and modal properties is presented. Detailed analysis of the temperature effect on the properties is presented. The Principal Component Analysis (PCA) methodology has been utilized to reduce some environmental effects on the vibration frequencies to identify earthquake damage in the structure. The advantages and disadvantages of the method are presented. Typical variations in the range of 4% are observed related to temperature and 6% related to the rain and the surrounding soil humidity. Earthquake damage during the 2010 Mw=8.8 Earthquake is clearly identify from ambient vibration and earthquake records. Damage is considered low, with some visible cracking in structural wall. Variations due to this damage are in the order for 15 to 20% for predominant natural frequencies.

15:10-15:30 — ThBT5.5

Identification of Inhomogeneous Concrete Cover by Non-Contact Ultrasonic Method

Piwakowski, Bogdan — Ec. Centrale de Lille

Liu, Qiang — Ec. Centrale de Lille (IEMN, TPIA UMR 8520 CNRS),
Balayssac, Jean-Paul — LMDC Toulouse
Villain, Géraldine — IFSTTAR
Garnier, Vincent — LMA CNRS
Kaczamrek, Mariusz — Kazimierz Wielki Univ. Bydgoszcz
Drelich, Radoslaw — Kazimierz Wielki Univ.

Concrete cover degradation is induced by aggressive agents in ambiance, chemicals, moisture, or temperature variations. Due to this degradation concrete becomes inhomogeneous and usually a thin surface layer appears with porosity and elastic modulus different than the properties of deeper sound concrete. Therefore the on-site nondestructive evaluation of concrete cover is important to monitor the integrity of concrete structures and prevent their irreversible damage. In this paper the methodology applied by the classical technique used for ground structure recovery called Multichannel Analysis of Surface Waves (MASW) is used as the NDT tool to characterize the thickness and elastic modulus of concrete cover. The procedure consists in generation and reception of surface waves within the required frequency band, using the non-contact transducers which scan evaluated profiles. Then the phase velocity dispersion characteristic V(f) is extracted, and concrete cover is characterized using as the proposed velocity gradient. The dispersion characteristics are exploited by the inversion software in order to obtain the variation of shear wave velocity as a function of depth.

ThBT6 — Room 200

Damage II (Regular Session)

Chair: Salvia, Michelle — Ec. Centrale de Lyon
Co-Chair: Villain, Géraldine — IFSTTAR

13:50-14:10 — ThBT6.1

Damage Identification in a Benchmark Cable-Stayed Bridge Using the Interpolation Method

Domaneschi, Marco — Pol. di milano
Limongelli, Maria Pina — Pol. di milano
Martinelli, Luca — Pol. di milano

In this paper the damage localization algorithm based on Operational Deformed Shapes (ODS) and known as Interpolation Damage Detection Method (IDDM), is applied to the numerical model of a cable stayed bridge. Frequency response functions (FRFs) have been calculated basing on the responses of the bridge to low intensity seismic excitation and used to recover the ODS both in the transversal and in the vertical direction. The analysis have been carried in the undamaged configuration and repeated in several different damaged configurations Results show that the method is able to provide the correct location of damage, provided the an accurate estimation of the ODSs is available.

14:10-14:30 — ThBT6.2

Impact Loads for Structural Damage Identification

Swiercz, Andrzej — Pol. Acad. of Sciences

In this paper we numerically study simple two dimensional frame structure loaded by impulse-like excitations. Various configurations of time duration of a single impact load and location of this excitation is investigated. The aim of this research is to indicate the best possible load case for identification of a specific damage. There are considered two kinds of a structural modification. The fist group concerns the element stiffness changes (axial and bending) which can be simulated by alternating the parameter of Young`s modulus. The second group covers the modifications of stiffness nodal connections. In a~numerical model, these kind of defects can be implemented using rotational spring with different characteristics. Obtained selected responses for original and modified structure is examined.

14:30-14:50 — ThBT6.3

Validation of Structural *Parameters As Damage Indicators for Monitoring Plates in the Post Buckling Regime*

Viechtbauer, Christoph — Johannes Kepler Univ. Linz
Schröder, Kai-Uwe — RWTH Aachen Univ.
Schagerl, Martin — Johannes Kepler Univ. Linz

The Structural Health Control (SHC) concept is a novel and overarching approach to monitor the strength of a structure online and in realtime over its lifecycle. In the center of the SHC approach is the structural analysis of the monitored part. It provides the information to identify structural parameters as damage indicators. With this information an optimized sensor layout can be defined to monitor the damage indicators in a smart way. The monitoring system is therefore referred as SmartSHM system. This article is about the experimental validation of the SmartSHM concept within the SHC approach. The SmartSHM system is tested at a simply supported thin plate under compressive loading close to ultimate load. In the post-buckling regime bending occurs which causes high tension stresses at the plate's surface despite of the compressive loading. These stresses can initiate a crack due to fatigue. Therefore these regions are monitored with a SmartSHM system. Cracks were initiated to prove the functionality of the proposed SHM concept.

14:50-15:10 — ThBT6.4

3D Elliptical Crack Depth Estimation from 2D Surface Displacement Observation

Pineau, Pierre — INSA Lyon
Rethore, Julien — LaMCoS, INSA Lyon, CNRS
Baietto, Marie-Christine — INSA UMRC 5259
Fregonese, Marion — INSA de Lyon

This paper presents a method to estimate the depth of an elliptical 3D crack from 2D observations on surface displacement fields. An iterative numerical procedure with leastsquares error minimisation is developped to compare pseudo-experimental displacements fields with a FEM displacement fields database and estimate crack depth.

15:10-15:30 — ThBT6.5

An Advanced Ensemble *Impact Monitoring and Identification Technique for* Aerospace *Composite Cantilever Structures*

Si, Liang — Tech. Univ. Munich
Baier, Horst — Tech. Univ. Munich

An investigation was performed to develop a real-time automatic health monitoring technique for the identification and prediction of the location and the force magnitude of foreign object impact on composite structures with distributed sensor network. In the smart ensemble impact identification (EII) technique proposed, it consists of four sequential procedures, which are the sensor signal preprocessing (SSP), the forward system modeling (FSM), the inverse model operator (IMO) and the impact positioning. Subsequently, in our experimental cases, we considered the disturbed factor -- random interfering noises, and added the cantilever support condition into our experimental tests of a CFRP plate structure, meanwhile, we also used the small balls with the different materials and masses as the impactors. However under the various impact situations and external noise environment, the predictions for the accuracy of impact forces and locations using the EII technique were validated, and the evaluated errors all fell well within the satisfactory limited range, and also interpreted the EII technique that is competent to reconstruct precisely the input-force signal due to stochastic impact event and estimate the impact location effectively in complex practical environment.

airbus.com
The A350 XWB.
Next generation efficiency with next level comfort.
• 25% lower operating costs than previous generation competitors
• All-new composite fuselage, highest operational reliability and outstanding aerodynamics
• Exceptional level of comfort with more personal space, wider seats and panoramic windows, inflight entertainment systems with full connectivity
Become part of the A350 XWB family, the most successful widebody aircraft of recent years.
AIRBUS

Friday July 11, 2014

Technical Program for

Friday July 11, 2014

Keynote Lectures	Auditorium
Chair: Mevel, Laurent	Inria

08:30-09:00	FrKN1L

SHM Trends and Opportunities *for the Civil Engineering Sector*

Bourquin, Frédéric — IFSTTAR

Godart, Bruno — IFSTTAR

The structures, especially bridges, are strategic elements of a network of road, railway or inland waterway infrastructures. They are constantly exposed to aggressive environments, to the aging of their equipments and their constituent materials, to the degradation of their structural elements and to changes in operating conditions. These latter are characterized by an increase of the traffic, by heavier loads and by more abnormal loads. These factors amplify the effect of potential errors in design or construction, or the consequences of accidental damages such as vehicle impacts.

The preservation of a long lasting patrimony of structures requires a management policy that includes not only actions of maintenance, repair, strengthening and rehabilitation, but also actions for limiting and even stopping the operation of structures, for installing a safety monitoring, or for immediate safeguarding. These actions are only relevant when appropriate procedures and methods exist to assess the residual load carrying capacity, the serviceability and durability of the structure.

Structural Health Monitoring is one of the important tools available to the manager of a patrimony in order to evaluate the structural performance of the bridges, or alternatively to have an objective and relevant vision of their condition. Referring to the S^3 project initiated by Ifsttar in 2007, Structural Health Monitoring can be divided into three themes:

- the monitoring by itself or operational monitoring which includes instrumentation, measurement and management of information;
- the diagnosis which consists of processing data collected during the monitoring phase (performance indicators, discrimination tests, alert thresholds,...);
- the structural identification whose objective is to provide predictive information on the structure behavior (resistance, life duration,...) which is integrated into the management system.

Although the structural health monitoring is not yet common practice, it is already applied for exceptional structures (Normandy Bridge, Millau Viaduct) or for strategic structures (dams, nuclear power plants,...) and it is developing for large structures. Weak maintenance budgets, often minimized to excess..., do not facilitate the deployment of this approach, but a significant growth in the market for structural health monitoring has still been felt in recent years.

However, recent advances in SHM concepts and technologies pave the ground to large-scale deployments by just providing increasingly affordable monitoring capabilities and services. Actually, multiscale monitoring, low(er) cost fiber optics, nanosensing construction materials or nanoelectronics as well as embedded computing for damage detection and SHM at large will change the perspective to effective deployment. Moreover, new markets for SHM technologies with higher pay-off expectations and more convincing RoI are emerging as potential key-drivers. From the theoretical point of view, potential large-scale deployments also raise new questions and opportunities at the frontier of state, data, and parameters.

The talk will report on a combined view of SHM trends and opportunities for transportation and other infrastructure networks.

09:00-09:30	FrKN2L

Validation, Verification *and Implementation of SHM at Airbus*

Paget, Christophe — Airbus

Bockenheimer, Clemens — Airbus

Structural Health Monitoring (SHM) is an innovative approach of Non-Destructive Testing (NDT) in order to ensure the integrity of aircraft structures. SHM links the world of structures and systems disciplines as the sensor remains permanently attached to the structure, i.e. both forming an entity. For this reason, the development of SHM technologies requires a highly transdisciplinary work, involving expertise from materials & processes and systems disciplines and taking into account numerous requirements from various transdisciplinary regulations and standards.

As a consequence, while the research and development of SHM is well established in industry and academia, it appears that the maturation towards reliable and applicable sensing technologies is still not satisfying from an end-user perspective. Hence, an adequate approach for the development and maturity assessment of SHM technologies needs to be realised.

The maturation of SHM technologies for a specific application scenario is closely linked to the Technology Readiness Level (TRL). Each of the TRL is covering a certain bundle of requirements. For a so called 'guided technology development' an appropriate allocation of requirements to TRL is essential in order to ensure a robust, time- and cost-efficient maturation of technologies. Within this maturation process a certain TRL level is reached if the relevant specific requirements have been fulfilled.

How can this be achieved? Firstly, for a given specific application scenario the requirements and their corresponding fulfillment criteria have to be defined. Then, it has to be proven that the selected requirements and fulfillment criteria are valid for the selected application scenario. In the verification phase, the fulfillment of selected requirements will be proven on the basis of variable analysis, comparison with conventional NDT, chemical and physical testing.

Starting from a look at the Airbus SHM roadmap in Airbus, the presentation will point out the way of working in Airbus to mature SHM technologies for selected application scenarios by covering the following scope: Definition of application scenarios, definition of the requirement families, definition of requirements and corresponding fulfillment criteria, validation and verification (V&V). This way of working is very much linked to the SAE Aerospace Recommended Practice ARP6461 'Guidelines for Implementation of Structural Health Monitoring on Fixed Wing Aircraft', which is the first guidance document for SHM in the field of Aerospace application.

Special attention will be paid to the V&V approach. A major role for the realisation of the V&V approach is played by the so called 'V&V Center for SHM' with the objective to provide the bases for a robust, practical, cost- and time efficient way in order to mature SHM technologies for selected application scenarios.

In conclusion, the presentation provides a valuable driving force as well as a guideline to those who are contributing or intending to contribute to the development of 'sensing structures'.

09:30-10:00	FrKN3L

Structural Monitoring *on Germanys Offshore Research Platform FINO3*

Jeromin, Andreas — FINO3

When Germany planned to put wind energy plants at offshore locations in the North Sea and the Baltic Sea, the need for additional information on the environmental conditions arose. Therefore, three different research platforms were installed on three remote locations where future offshore wind farm projects were to be realized. The third one is FINO3, located in the North Sea about 80 km off the island of Sylt. In this shallow water region the wind farm Dan Tysk is currently under construction.

This research platform is built on a monopile foundation and provides a working platform of 13 x 13 m² in size, a measurement mast with a top level of about 100 m above mean sea level and a helicopter platform for access of scientific and technical personnel. The research conducted ranges from meteorological and sea state measurements to biological observations and studying offshore foundations and structural performance.

In this study the differences between the loads from design assumptions and the actual loads on the existing structure are to be analyzed. Therefore, a monitoring system was installed on FINO3 that records the acceleration at three different heights. From these the displacement of the structure is computed and analyzed with respect to Eigen frequencies and Eigen modes. The displacement also enables us to model the platform and to find the local stress factors at critical parts, for example at welds, holes or mountings. With the local stress factors a real load spectrum is available for a durability analysis.

With FINO3 we have access to a well-equipped and public research platform where we gain insight into the structural characteristics of the monopile. The monopile is also free from loads of an operating wind turbine and is only affected by the environment.

FrAT1 Auditorium

Structural Identification and Damage Detection Using Vibration Measurements (Invited Session)

Chair: Papadimitriou, Costas	Univ. of Thessaly
Co-Chair: Chatzi, Eleni	Inst. of Structural Engineering, ETH Zurich

10:30-10:50 FrAT1.1

Precise Vibration-Based Damage Localization in 3D Structures Consisting of 1D Elements: Single vs Multiple Response Measurements (I)

Sakaris, Christos	Univ. of Patras
Sakellariou, John	Univ. of Patras
Fassois, Spilios D.	Univ. of Patras

The goal of this study is twofold: (i) the refinement of the advanced Functional Model Based Method (FMBM), which through a detailed representation of 3D structures consisting of 1D elements achieves precise damage localization based on single or multiple vibration responses and, (ii) the comparison of the method's effectiveness based on single and multiple response measurements. The refined method is equipped with modified ARX type Functional Models - Vector Functionally Pooled Vector AutoRegressive with eXogenous excitation (VFP-VARX) models - for the simultaneous exploitation of multiple responses and a proper optimization framework based on which the precise estimation of the damage coordinates upon the 3D structural topology under study is accomplished. The method's effectiveness and the comparison between the use of single or multiple response signals are experimentally assessed via numerous damage cases in a 3D truss structure.

10:50-11:10 FrAT1.2

Simple Techniques to Analyze *Vibration Records from Multi-Story Buildings (I)*

Safak, Erdal	Bogazici Univ. Kandilli Observatory and Earthquake Res.
Cakti, Eser	Bogazici Univ. Kandilli Observatory and Earthquake Res.

In parallel to rapid developments in recording, sensor, and communication technologies, the number of structures installed with vibration monitoring systems is increasing rapidly all over the world. The scientific field dealing with the analyses of structural monitoring data is known as System Identification, and involves a large number of advanced techniques requiring expertise and background for usage. Since analysis of digital data is typically not covered in standart structural engineering curriculum, a large amount of data from monitored structures are not being analyzed properly and correctly. This paper shows that there are simple techniques to analyze vibration data from structures. For building-type structures, it is possible to identify basic dynamic characteristics of a structure by using only Fourier transforms and band-pass filters. We show that, by using these two tools, we can identify modal frequencies and damping ratios, mode shapes, interstory drifts, torsional vibrations, rocking vibrations, and soil-structure interaction.

11:10-11:30 FrAT1.3

Dispersion-Corrected Stabilization Diagrams for Model Order Assessment in Structural *Identification (I)*

Dertimanis, Vasilis	Inst. of Structural Engineering, ETH Zürich
Chatzi, Eleni	Inst. of Structural Engineering, ETH Zurich

This study aims at overcoming some of the inconsistencies of stabilization diagrams, by applying a novel dispersion analysis framework that relies on the effective modal decomposition of the vibration output's zero-lag covariance matrix, under the assumption of broadband random excitation. A new metric is introduced, which expresses a certain part of the total stochastic vibration energy, and is attributed to each vibration mode. It is then shown how this metric can be integrated into a stabilization diagram, in order to enrich the displayed information and facilitate the model order selection process. In this respect, a vibration mode is identified as structural when it appears stabilized in both its frequency and its dispersion metric. The method's performance is assessed through structural identification problems from both simulated and experimental vibration data, for which subspace and prediction-error based methods are utilized.

11:30-11:50 FrAT1.4

Identification and Damage Detection of a Shear Frame Model Based on a Blind Source *Separation Method (I)*

Spiridonakos, Minas	ETH Zurich
Yadav, Nishant	Univ. of Michigan Ann Arbor
Chatzi, Eleni	Inst. of Structural Engineering, ETH Zurich

Blind source separation methods have recently emerged as a competitive alternate structural identification method offering a number of advantages when compared to traditional methods. These include the plain underlying model, its computational ease, and its robustness to various types of excitations. The aim of this study, is to experimentally assess the effectiveness of a second order BSS method for the problems of output-only structural identification and damage detection, by its application on the vibration response data of a four-storey shear frame. The structure under study is subjected to various types of input excitation and is tested under its ``healthy" and a number of simulated ``damaged" states. The results of the study show great potential of the BSS method for the efficient treatment of both problems.

11:50-12:10 FrAT1.5

Uncertainty Calibration *of Large-Order Models of Bridges Using Ambient Vibration Measurements (I)*

Papadimitriou, Costas	Univ. of Thessaly
Argyris, Costas	Univ. of Thessaly
Papadioti, Dimitra-Christina	Univ. of Thessaly
Panetsos, Panagiotis	Egnatia Odos S.A.

A computational efficient Bayesian inference framework based on stochastic simulation algorithms is presented for calibrating the parameters of large-order linear finite element (FE) models of bridges. The effectiveness of stochastic simulation tools to handle large-order linear models in Bayesian analysis is demonstrated by calibrating a high fidelity FE model of the Metsovo bridge with several hundreds of thousands of DOF, using experimentally identified modal frequencies and mode shapes based on ambient vibration measurements collected from a wireless mobile measuring system. The mode shapes of the bridge are assembled using the identified modal characteristics from a number of different sensor configurations, involving reference and moving sensors, optimally placed on the bridge deck to adequately cover the whole bridge span. The identified finite element models and their uncertainties are representative of the initial structural condition of the bridge and can be further used for structural health monitoring purposes.

12:10-12:30 FrAT1.6

Vibration-Based Structural Health Monitoring of High Voltage Electrical Switch Ceramic Insulators *in Seismic Areas*

Rébillat, Marc — PIMM, Arts et Métiers ParisTech
Barthes, Clement — UC Berkeley -PEER Lab.
Mechbal, Nazih — Arts et métiers ParisTech
Mosalam, Khalid — UC Berkeley

High voltage electrical switches are crucial components to restart rapidly the electrical network right after an earthquake. But there currently exists no automatic procedure to check if these ceramic insulators have suffered after an earthquake, and there exists no method to recertify a given switch. To deploy a vibration-based structural health monitoring method on ceramic insulators a large shake table able to generate accelerations up to 3 g was used. The idea underlying the SHM procedure proposed here is to monitor the apparition of cracks in the ceramic insulators at their early stage through the change of the resonant frequency of the first mode of the structure and the non-linearity that they generate in its dynamic response. The Exponential Sine Sweep Method is used to estimate a nonlinear model of the structure under test from only one dynamic measurement. A classic linear damage index (DI) based on the variation of the frequency of the first mode is compared to an original nonlinear one using the ratio of the amplitudes of the third harmonic and the fundamental frequency. Results show that both DIs increase monotonically with the number of solicitations, thus validating the use of the nonlinear DI. It is also shown that the nonlinear DI presented here seems more sensitive than the linear one.

FrAT2 — GH Room
Composites II (Regular Session)

Chair: Lecce, Leonardo — Department of Industrial Engineering - Univ. of Naples Federico II

Co-Chair: Le Gal La Salle, Eric — ICAM

10:30-10:50 FrAT2.1

Multiscale Model and Experimental Study of Damage in Piezoelectric Fiber-Based Composites

Shalan, Khalid — The British Univ. in Egypt
Abdel-Meguid, Mohamed — The British Univ. in Egypt
Hatem, Tarek — The British Univ. in Egypt
Bahei-El-Din, Yehia — The British Univ. in Egypt

The ability of piezoelectric materials to couple mechanical and electrical properties have caused the materials to play a pioneer role in structural health monitoring (SHM) applications as well as energy harvesting systems. Piezoelectric materials are able to perform the role of both sensors and actuators in SHM systems. However, the electromechanical behavior of Piezoelectric Fiber Composites (PFCs) remains a viable field of study. Micromechanical model based on transformation field analysis is used to quantify the overall material properties of electrically active composite structure. Capitalizing on the extracted properties, single-phase analysis of a homogeneous structure is conducted using finite element method, extended to include electric behavior of active materials. A numerical modal analysis based on the aforementioned multiscale model is conducted on pristine samples. An experimental study focuses on PFC damage such as delaminations could impair its role as a sensor or actuator in an SHM system or in an energy harvesting system. An experimental modal analysis is performed on piezoelectric fiber based composite bimorphs in both pristine and damaged conditions. Shifts in natural frequencies and the corresponding electrical response at resonance due to a delamination introduced to a PFC.

10:50-11:10 FrAT2.2

Laser Ultrasonic Imaging for Impact Damage Visualization in Composite Structure

Zhang, Chao — Nanjing Univ. of Aeronautics and Astronautics
Qiu, Jinhao — Nanjing Univ. of Aeronautics and Astronautics
Ji, Hongli — Nanjing Univ. of Aeronautics and Astronautics

Laser ultrasonic scanning technique has great potential for damage evaluation in various applications. In order to detect the size and shape of the damage from the wave propagation information, this paper presents an improved imaging method based on the anomalous incident wave (AIW) energy. Compared with the original wavefield, the AIW filters the reflected wave by using wavenumber-frequency domain analysis and eliminates the traveling waves by means of adjacent incident waves subtraction. Taking these advantages of proposed method, the changes in wave energy distribution caused by the damage can be highlighted to show the size and shape of the damage. Finally, a carbon fibre reinforced polymer (CFRP) laminated plate with an impact damage is used to validate the proposed method. From the result, the proposed method improves the resolution of the damage evaluation.

11:10-11:30 FrAT2.3

CFRP Structural *Health Monitoring by Ultrasonic Phased Array Technique*

Boichuk, Alexander — FSUE "VIAM"
Generalov, Alexander — FSUE "VIAM"
Stepanov, Alexander — FSUE "VIAM"

The report deals with ultrasonic phased array (PA) application for high-loaded CFRP structural health monitoring in aviation. Principles of phased array technique and most dangerous types of damages are briefly described. High-performance inspection technology suitable for periodic plane structure check is suggested. The results of numerical estimation of detection probability for impact damages and delaminations by PA technique are presented. The experience of PA implementation for designing and efficiency estimation of integrated CFRP structural health monitoring system based on FBGA for impact damage detection is described.

11:30-11:50 FrAT2.4

Delamination Detection in Fiber Metal Laminates Using the Mode Conversion of Lamb Waves

Okabe, Yoji — Univ. of Tokyo
Hirakawa, Hibiki — Tokyo Univ. of Science
Nakatani, Hayato — Osaka City Univ.
Ogihara, Shinji — Tokyo Univ. of Science

Fiber metal laminates (FMLs) consisting of FRP composites and thin metal foils are gathering attention as structural materials for aircrafts. However, it is difficult to apply the conventional ultrasonic method to detect the inside damages because of the total reflection at the first interface between the FRP and the metal. Recently, the authors have developed a new delamination detection method based on the mode conversion of Lamb waves. In this research, therefore, this method was applied to detect the inner delamination in the FML. MFC actuators and FBG sensors were bonded to both top and bottom surfaces of the laminate in order to excite or receive only A modes or S modes separately. The laminates with an artificial delamination were manufactured by embedment of Teflon films in the middle of the thickness, and Lamb waves were propagated through the delaminated area. The results indicated that excited A1 mode was converted into S0 mode in the delaminated area and it returned to A1 mode again after passing through the delamination. Since the velocity dispersion is clearly different between the two modes, the delamination length could be estimated quantitatively from the change in the dispersion of received A1 mode.

11:50-12:10 FrAT2.5

Quasi-Continuous Mode Conversion of Lamb Waves in CFRP Plates Due to Inhomogeneity *on Micro and Meso Scale*

Neumann, Mirko	Helmut-Schmidt-Univ. der Bundeswehr Hamburg
Hennings, Bianca	Helmut-Schmidt-Univ.
Lammering, Rolf	Helmut-Schmidt-Univ. der Bundeswehr Hamburg

The mode conversion of symmetric into antisymmetric Lamb waves in non-damaged CFRP plates caused by their inhomogeneity is discussed. Lamb waves in differently stacked CFRP plates are observed via scanning laser vibrometry. In addition to the dominating wave crests observed propagating through the plates, as predictable from a macro-mechanical point of view, different patterns of parallel lines occur. These are analyzed and classified into three effects: Mode conversion at micro scale (fibers) and a static bending-stretching coupling effect at meso scale (weave rovings), also causing mode conversion under additional conditions. To consolidate the findings, tensile tests are performed and observed with a 3D image correlation method and a volume FE model of a twill fabric woven layer is used for static as well as dynamic simulations. The numerical results and the experimental data show a good agreement.

FrAT3 Room I

SHM Applications III (Regular Session)

Chair: Guemes, Alfredo	UPM
Co-Chair: Döhler, Michael	Inria

10:30-10:50 FrAT3.1

The British Columbia Smart Infrastructure Monitoring System (BCSIMS)

Kaya, Yavuz	The Univ. of British Columbia
Turek, Martin	The Univ. of British Columbia
Huffman, Sharlie	BC Ministry of Transportation & Infrastructure
Ventura, Carlos	The Univ. of British Columbia

The British Columbia Ministry of Transportation and the University of British Columbia have implemented a program to instrument key structures to provide confirmation of seismic capacity, assist in focusing retrofit efforts, perform structural health evaluations and provide rapid damage assessment of those structures following a seismic event. The instrumentation system installed at each structure will automatically process and upload data to a central server via the Internet. The alert systems and public-access web pages can display real time seismic data from the structures and from the BC Strong Motion Network to provide input for assessments by the Ministry of non-instrumented bridges. These systems may also provide other agencies, emergency responders and engineers with situational awareness.

10:50-11:10 FrAT3.2

A Highly Instrumented Underground Research Gallery As a Monitoring Concept for Radioactive Waste Cells - Data Measurement Qualification

Farhoud, Radwan	ANDRA
Hermand, Guillaume	ANDRA
Kazmierczak, Jean-Bernard	INERIS
Malherbe, Laure	INERIS
Balland, Cyrille	INERIS

This document presents a dedicated work performed in Andra's underground research laboratory (URL) in order to prepare, to test and to qualify the future monitoring system of underground disposal for the French long-lived, intermediate and high level radioactive wastes. Part of the monitoring system qualification process relies on testing sensors on full-scale demonstrators and ensuring that we carefully observe the desired parameters. One of these demonstrators is a concrete liner in a tunnel aiming at support the mechanical pressure of the host rock. A 3.6 meter long section of this gallery has been highly instrumented by various technologies of sensors. This paper describes the monitoring system installed and demonstrates how a numerical model of the demonstrator has been utilized to comfort and validate sensors' measurements.

11:10-11:30 FrAT3.3

Quantification and Uncertainty Analysis of a Structural Monitoring Device : Application to the Detection of Chloride in Concrete Using Electrical Resistivity

Lecieux, Yann	GeM, Univ. of Nantes
Schoefs, Franck	Univ. of Nantes
Bonnet, Stéphanie	GeM Univ. of Nantes
Lopes, Sergio Palma	IFSTTAR
Roche, Michel	GeM, Univ. of Nantes

In this work, we seek to assess and optimize the performances of an integrated resistivity sensor compatible with Geoelectrical Imaging methods. In this way, we have performed tests in a controlled environment. We specifically seek to evaluate the detection threshold of chlorides by doing tests on concrete specimens containing different concentrations of these ions. Performance assessment of the resistivity probe is based on an analysis of ROC curves and the detection threshold of chlorides is calculated using the alpha-delta method.

11:30-11:50 FrAT3.4

3D Printing for Intelligent *Metallic Structures*

Strantza, Maria	Vrije Univ. Brussel
De Baere, Dieter	Vrije Univ. Brussel
Rombouts, Marleen	VITO Vlaamse Inst. voor Tech. Onderzoek
Clijsters, Stijn	Katholieke Univ. Leuven
Vandendael, Isabelle	Vrije Univ. Brussel
Terryn, Herman	Vrije Univ. Brussel
Guillaume, Patrick	Vrije Univ. Brussel
Van Hemelrijck, Danny	Vrije Univ. Brussel

Structural Health Monitoring (SHM) is needed both to improve life-safety and to reduce the direct operational costs. Nowadays, a new concept called effective Structural Health Monitoring (eSHM) has proven to be efficient but still needs to pass different technological readiness levels to enable its implementation. This work demonstrates the feasibility study of eSHM systems produced by 3D printing or additive manufacturing. The objective of this work is to prove that the eSHM system has reached technological readiness level 3 (TRL3) and to indicate that during fatigue the integrated system had no influence on the crack initiation behaviour. First, two different techniques (selective laser melting and laser metal deposition) were used for the production of four-point bending feasibility test specimens with the integrated eSHM system. Next, a four-point bending test was selected; the specimens were subjected to the so-called step test method with constant fatigue stress amplitude and a constant R ratio. The fatigue behaviour of stainless steel and titanium alloy was studied with emphasis on crack initiation and detection. This study proves that the eSHM reaches the TRL3.The investigated system always detected cracks although further investigation is needed since higher TRL are required and the detection capability can be improved.

11:50-12:10 FrAT3.5

Path Planning & Measurement Registration for Robotic Structural Asset Monitoring

Pierce, Stephen Gareth	Univ. OF STRATHCLYDE
Macleod, Charles Norman	Centre for Ultrasonic Engineering (CUE), Department of Electroni
Dobie, Gordon	Univ. of Strathclyde
Summan, Rahul	Univ. of Strathclyde

The move to increased levels of autonomy for robotic delivery of inspection for asset monitoring, demands a structured approach to path planning and measurement data presentation that greatly surpasses the more ad-hoc approach typically employed by remotely controlled, but manually driven robotic inspection vehicles. The authors describe a traditional CAD/CAM approach to motion planning (as used in machine tool operation) which has numerous benefits including the provision for detailed structural layout to be built into the path planning, and the subsequent visual overlay of measurement data back into the CAD model. Initial trials using a steel plate with varying thickness defects have been investigated. A custom robot incorporating an ultrasonic wheel probe as rear wheel has been used to perform an ultrasonic thickness scan of the sample – the results have been compared to those obtained using a high resolution wheel probe scan of the sample. The robot control was generated from an initial CAD model of the sample, and the integration of post-processor output (in this case G code), was used as the input for our custom robotic control software to generate both simulation and experimental motion control of our robotic platform.

FrAT5 KL Room

Signal Processing (Regular Session)

Chair: Ostachowicz, Wiesław — Pol. Acad. of Sciences, Inst. of Fluid Flow Machinery

Co-Chair: Sakellariou, John. — Univ. of Patras

10:30-10:50 FrAT5.1

On the Stability of Sequential *Deconvolution*

Bernal, Dionisio — Northeastern Univ.

The time domain estimation of inputs from knowledge of the kernel and the outputs is a de-convolution operation. For finite dimensional systems treated in discrete time the operation is tantamount to solving a set of linear equations. When performing a deconvolution an issue that must be dealt with is the fact that the dimension of the system of equations grows with duration and becomes prohibitively large when the inputs are long. For this reason, and sometimes because it is of interest to estimate the inputs with the smallest possible delay, deconvolution must often be implemented on a moving window. This paper shows that a sequential deconvolution is a conditionally stable process and derives the expression that governs numerical stability.

10:50-11:10 FrAT5.2

Ensemble Empirical Mode Decomposition (EEMD) and Teager-Kaiser Energy Operator (TKEO) Based Damage Identification of Roller Bearings Using One-Class Support Vector Machine

Tabrizi, Ali — Politecnico di Torino
Garibaldi, Luigi — Politecnico di Torino
Fasana, Alessandro — Politecnico di Torino
Marchesielo, Stefano — Politecnico di Torino

Ensemble empirical mode decomposition (EEMD) is a newly developed noise assisted method aimed to solve mode mixing problem exists in empirical mode decomposition (EMD) method. Although EEMD has been utilized in various applications successfully, small defects of bearings are not able to be detected, especially in automatic defect detection, when only healthy samples are available for training. Teager-Kaiser energy operator (TKEO) technique is a non-linear operator that can track the energy and identify the instantaneous frequencies and instantaneous amplitudes of signals at any instant. As Teager-Kaiser energy operator (TKEO) technique detects a sudden change of the energy stream without any priori assumption of the data structure, it can be utilized for vibration based condition monitoring (non-stationary signals). In this study it is investigated whether an automatic method is able to diagnose a small defect level of roller bearings through processing of the acquired signals. After applying TKEO on IMFs decomposed by means of EEMD, the extracted informative feature vectors of the healthy bearing are used to construct the separating hyperplane using one-class support vector machine (SVM). Then, success rates of state identification of both samples (healthy and faulty) are examined by labelling the samples. The data were generated by means of a test rig assembled in the labs of the Dynamics & Identification Research Group (DIRG) at mechanical and aerospace engineering department, Politecnico di Torino. Various operating conditions (three shaft speeds, three external loads and one small size damage on a roller) were considered to obtain reliable results.

11:10-11:30 FrAT5.3

Structural Health Monitoring *in a Buckled Beam Using Volterra Series*

Hansen, Cristian — Unesp - Univ. Estadual Paulista
Da Silva, Samuel — UNESP - Univ. Estadual Paulista

This paper proposes a new method to detect damages in structures vibrating with nonlinear behavior. The approach proposed is based on Volterra series and can separate the linear and nonlinear contributions of the responses. An alert of damage is given based on level of contribution of the difference between the linear and nonlinear behavior identified using experimental time-series. Tests are performed involving a buckled clamped beam to illustrate the steps and the advantages of the approach to detect damages using vibrating in nonlinear regime of motion.

11:30-11:50 FrAT5.4

Phase Statistics and Spectral Analysis of Ultrasonic Signals for CFRP Component Assessment

Lozak, Alexander — IDEAL Tech.
Boller, Christian — Saarland Univ.
Bulavinov, Andrey — IDEAL Tech.
Pinchuk, Roman — IDEAL Tech.
Kurz, Jochen — Fraunhofer IZFP
Sednev, Dimitry — IDEAL Tech.

The established and industrially applied ultrasonic inspection and monitoring techniques typically use the amplitude values of the ultrasonic signals for assessment of flaws. However every individual signal value has the phase information, in addition to the amplitude information. Through analysis of the statistical distribution of phase values provided by state-of-the-art measurement systems and an appropriate consideration of this information in the signal evaluation, a significant improvement in flaw detection capability of an ultrasonic test can be achieved. Moreover, the phase and the frequency of ultrasonic signals carries information that can be correlated to the inner structure and existing inhomogenities of the material to be assessed. The two novel approaches for increasing the information content of the classic ultrasonic test - phase statistics and spectral analysis - are new and promising. Use of these approaches, even for on-line measurements has become possible particularly through modern computer science and signal processing. It is particularly significant that these methods are based exclusively on the analysis and processing of "conventionally" acquired ultrasonic signals, so that in-depth rework of the measurement technology or ultrasonic hardware can be added with additional useful information on the material's condition to be obtained.

11:50-12:10 FrAT5.5

A Signal Processing Method for Hits Detection and Separation in High AE Activity Systems*: Application to Composite Materials under Fatigue Tests*

Kharrat, Mohamed — Femto-st Inst.
Ramasso, Emmanuel — Femto-st Inst.
Placet, Vincent — Femto-st Inst.
Boubakar, Mohamed Lamine — Femto-st Inst.

The threshold-based technique, employed by most of the commercial parameter-based AE systems for hit detection, uses a defined threshold in order to detect the start and the end of the hits. When dealing with continuous emission, the threshold-based technique is not suitable as the burst never drop below the threshold. In this case, the AE system is obliged to force the end of the hit after a defined maximum duration. This issue is encountered in complex systems such as in Organic Matrix Composites (OMC) fatigue tests where a high AE activity is observed, especially when the loading frequency overpasses few hertz. In this case, continuous signals can be generated by the background noise and rubbing, as well as by both cumulated damage (friction of crack surfaces) and damage growth. This paper deals with continuous AE signals obtained from experimental tests carried on Carbon Fiber Reinforced Plastics specimens. A numerical routine was implemented allowing the treatment of these signals. As the size of each acquisition is large due to the sampling rate (generally from 2 to 5 MS/s), the signal was divided into short segments. The Discrete Wavelet Transform (DWT) was used for signal denoising after adapting a certain number of parameters. Hit detection and determination was thereafter performed in order to localize potential hits contained in each signal segment. Conventional AE features were then calculated. By comparing the obtained results to those of conventional threshold-based techniques, we remark that the problem of erroneous hits is overcome. AE information that was hidden by the effect of the noise is now revealed allowing a further interpretation of damage mechanisms in the composite.

FrAT6 — Room 200
SHM Methodologies (Regular Session)

Chair: Staszewski, Wieslaw J.	Department of Robotics and Mechatronics, AGH Univ. of Science and Technology, Kraków
Co-Chair: Holnicki-Szulc, Jan	IFTR Warsaw

10:30-10:50 — FrAT6.1

Scheduled and SHM Structural Airframe Maintenance Applications Using a New Probabilistic Model

Cot, Léa Dominique	Inst. Clément Ader
Wang, Yiwei	Inst. Clément Ader, Univ. of Toulouse ; INSA, UPS, Mines
Bes, Christian	Inst. Clément Ader
Gogu, Christian	Univ. of Toulouse

This paper focuses on aircraft structure maintenance. A general mathematical framework based on a probabilistic analytical model was developed to provide the average number of fuselage panels to be replaced after any number of cycles and over the aircraft lifetime under fatigue damage failure. This paper proposes a study to explore structure maintenance strategies using this analytical model. Both scheduled maintenance and condition-based maintenance using SHM are considered for different aircrafts with different lifetimes and timetable inspections. In each case, the aircraft lifecycle cost is evaluated in terms of average number of panels changed. A similar study based on Monte Carlo method was carried out and we studied the results given by this approach considering same maintenance strategies as mentioned above. Because the simulation conditions are not exactly the same, the results obtained using the PAM are a little bit different than the ones obtained by Monte Carlo simulations; indeed, it shows an adequacy. Compared to Monte Carlo method, the analytical model is of a great interest regarding the accuracy because the recurrence formula leads to an exact expression of the average number of fuselage panels to be replaced.

10:50-11:10 — FrAT6.2

Effective Structural Health Monitoring *with Additive Manufacturing*

De Baere, Dieter	Vrije Univ. Brussel
Strantza, Maria	Vrije Univ. Brussel
Hinderdael, Michaël	Vrije Univ. Brussel
Devesse, Wim	Vrije Univ. Brussel
Guillaume, Patrick	Vrije Univ. Brussel

The current remaining challenges for structural health monitoring (SHM) systems prevented the introduction of SHM systems on a large scale within industrial applications. An effective SHM system is required for the reduction of the direct operation costs, improvement of the life-safety and the introduction of additive manufactured components for critical structures. In this paper, a new structural health monitoring methodology will be presented for components that can be processed by additive manufacturing (AM) or 3D printing. The origin of the source concept will be explained together with the different possible configurations and the intrinsic strengths of the system. It will be concluded that these new AM production techniques and new SHM technique will allow the development of "intelligent" metal structures in the future. The importance of the SHM system for the breakthrough of the AM techniques will be highlighted for critical components such as primary aeronautical structures.

11:10-11:30 — FrAT6.3

On the Damage Diagnosis Based on Structural Analysis Data

Schröder, Kai-Uwe	RWTH Aachen Univ.
Preisler, Andreas	RWTH Aachen Univ.
Viechtbauer, Christoph	Johannes Kepler Univ. Linz
Schagerl, Martin	Johannes Kepler Univ. Linz

One of the major challenges of Structural Health Monitoring (SHM) is to handle the enormous amount of data during the monitoring action. This is due to the fact that common SHM approaches monitor the entire structure on all kind of damages in order to obtain all eventualities of damage. The idea of this contribution is to use the information of the part's sizing in order to identify so called hot spots, i.e. possible damage locations and the corresponding kind of damage like cracks or delaminations. With this information a SmartSHM system is set up, which monitors only the hot spots on the most likely failure. To cover uncertainties and accidental damages an additional monitoring of a global parameter has to be set up. This approach is shown based on an example of a four point bending test with a rectangular cross-section.

11:30-11:50 — FrAT6.4

Requirements to Establish Fibre-Optic Sensors for Monitoring of Structures

Habel, Wolfgang R.	BAM Federal Inst. for Materials Res. and Testing
Schukar, Vivien Gisela	BAM Federal Inst. for Materials Res. and Testing
Hofmann, Detlef	BAM Federal Inst. for Materials Res. and Testing

Fibre-optic sensors need to be more established in the sensor market. Their advantages have unquestionably been verified by numerous demonstrations. However, there are some open questions leading now and then to restraints in the user's community. The paper discusses examples where fibre-optic sensors provide outstanding knowledge about the structure's behaviour, but application is often challenging. Requirements are listed and open questions to be solved are discussed. Short outlook to standards useful for better design, characterization and application is given. Standards are the basis for establishing SHM systems, especially for safety-critical structural diagnostics.

11:50-12:10 — FrAT6.5

Frequency Domain Instantaneous *Wavenumber Estimation for Damage Quantification in Layered Plate Structures*

Mesnil, Olivier	Georgia Inst. of Tech.
Yan, Hao	Georgia Inst. of Tech.
Ruzzene, Massimo	Georgia Inst. of Tech.
Paynabar, Kamran	Georgia Inst. of Tech.
Shi, Jianjun	Georgia Inst. of Tech.

Guided wavefield detection is at the basis of a number of promising techniques for the identification and the characterization of damage in plate structures. Among the processing techniques proposed, the estimation of instantaneous wavenumbers can be used as an effective metric that localize and quantifies delaminations in composite plates. A process able to estimate the in-plane and out-of-plane (depth) coordinate of a feature in a 2D structure using the Frequency Domain Instantaneous Wavenumber (FDIW) damage quantification technique is detailed in this paper. A post processing algorithm using a smooth sparse decomposition is used to highlight the studied features. The effectiveness of this method combined to the post processing technique is demonstrated for both numerical and experimental cases. This proposed methodology can be considered as a first step towards a hybrid structural health monitoring/ nondestructive evaluation (SHM/NDE) approach for damage assessment in composites.

12:10-12:30 FrAT6.6

Structural Parameters Identification *Using FRF of Incomplete Strain Data*

Shadan, Fariba — Amirkabir Univ. of Tech.
Esfandiari, Akbar — Amirkabir Univ. of Tech.
Khoshnoudian, Faramarz — Amirkabir Univ. of Tech.
Pedram, Masoud — Amirkabir Univ. of Tech.

Structural health monitoring by estimation of structural parameters changes such as stiffness and mass parameters by monitoring dynamic characteristics has attracted much attention in recent decades. Generally used dynamic characteristics to locate and quantify structural damages are natural frequencies, mode shapes, mode shape curvature, modal strain energy, frequency response function (FRFs) and so forth. Using FRF data has the advantage of avoiding modal analysis errors included due to indirectly extraction from the measured FRF data. Moreover, past studies showed that strains are more sensitive to localized damage compare to displacement. So, in this study FRF of strain data are utilized to identify unknown structural parameters using a sensitivity-based model updating approach. In this paper a quasi-linear sensitivity equation which diminishes the adverse effects of incompleteness of FRFs data is proposed for model updating. The efficiency of the proposed method is validated through a numerical 2D-frame example using FRF of strain data considering the side effects of noise and incompleteness of measurements. The results indicate that this method can locate and quantify the severity of damage precisely.